Study Guide with Selected Solutions
for David Moore's

The Basic Practice of Statistics
Fourth Edition

Michael A. Fligner
William I. Notz
The Ohio State University

W. H. Freeman and Company
New York

ISBN-13: 978-0-7167-7725-0
ISBN-10: 0-7167-7725-8

Printed in the United States of America

First printing

W. H. Freeman and Company
41 Madison Avenue
New York, NY 10010
Houndmills, Basingstoke RG21 6XS England

www.whfreeman.com

CONTENTS

CHAPTER 1

PICTURING DISTRIBUTIONS WITH GRAPHS

OVERVIEW

Understanding data is one of the basic goals in statistics. To begin, identify the **individuals** or objects described, then the **variables** or characteristics being measured. Once the variables are identified, you need to determine whether they are **categorical** (the variable puts individuals into one of several groups) or **quantitative** (the variable takes meaningful numerical values for which arithmetic operations make sense). The guided solution for Exercise 1.1 provides more details on deciding whether a variable is categorical or quantitative.

After looking over the data and digesting the story behind it, the next step is to describe the data with graphs. Simple graphs give the overall pattern of the data. Which graphs are appropriate depends on whether or not the data are numerical. Categorical data (nonnumerical data) are graphed in **bar charts** or **pie charts**. Quantitative data (numerical data) are graphed in **histograms** or **stemplots**. Quantitative data collected over time use a **time plot** in addition to a histogram or stemplot.

When examining graphs, be on the alert for the following:

- **Outliers** (unusual values) that do not follow the pattern of the rest of the data

- Some sense of a **center** or typical value of the data

- Some sense of how **spread** out or variable the data are

- Some sense of the **shape** of the overall pattern

In time plots, be on the lookout for **trends** over time. These features are important whether we draw the graphs ourselves or depend on a computer to draw them for us.

GUIDED SOLUTIONS

Exercise 1.1

KEY CONCEPTS: Individuals and types of variables

(a) When identifying the individual or objects described, you need to include sufficient detail so that it is clear which individuals are contained in the data set.

(b) Recall that the variables are the characteristics of the individuals. Once the variables are identified, you need to determine whether they are categorical (the variable puts individuals into one of several groups) or quantitative (the variable takes meaningful numerical values for which arithmetic operations make sense). Now list the variables recorded and classify each as categorical or quantitative.

Name of variable   Type of variable

Exercise 1.3

KEY CONCEPTS: Drawing bar charts

(a) What is the total of the percents in the table? Use the total to compute the percent of vehicles that are some other color.

(b) Complete the followint bar chart. The first bar has been drawn for you. Would a pie chart be appropriate if you added an "Other" category? What about without an "Other" category?

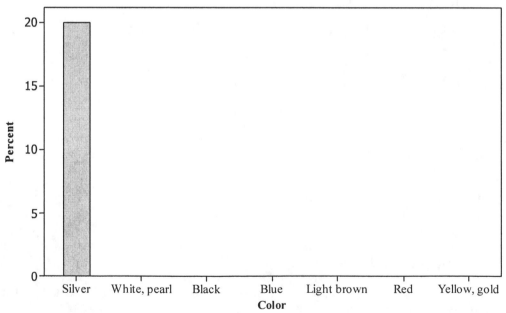

Exercise 1.11

KEY CONCEPTS: Drawing stemplots, splitting stems and rounding

Hints for drawing a stemplot:

1. It is easiest, although not necessary, to first order the data. If the data have been ordered, the leaves on the stems will be in increasing order. The fasting plasma glucose levels are ordered here.

```
Glucose
    78    95    96   103   112   134   141   145   147   148   153   158   172
   172   200   255   271   359
```

2. Decide how the stems will be shown. Commonly, a stem is all digits except the rightmost. The leaf is then the rightmost digit. Since a stemplot of these data would have many stems and no leaves or just one leaf on most stems, we first **round** the data to the nearest 10 mg/dL. The rounded data follow.

```
Glucose
    80   100   100   100   110   130   140   150   150   150   150   160   170
   170   200   260   270   360
```

3. Write the stems in increasing order vertically. Write each stem only once, unless you are splitting the stems. In this case, using the stems as the first digits (100s) would result in the leaves all falling on just a few stems. Because of this, it is best to split the stems. Rounding and splitting are matters of judgment, similar to choosing the classes in a histogram. Now draw a vertical line next to the stems. Write each leaf (10s column) next to its stem in the plot above. We have included the smallest and largest observations in the partial stem plot below to help you get started.

```
0 | 8
1 |
1 |
2 |
2 |
3 | 6
```

To finish up the exercise, think about the important features that describe a distribution. Does the distribution of the glucose measurements have a single peak? Does the distribution appear to be symmetric, or is it skewed to the right (tail with larger values is longer) or to the left? Are there any outliers that fall outside the overall pattern of the data? How well is the group as a whole achieving the goal for controlling glucose levels?

Exercise 1.33

KEY CONCEPTS: Drawing a histogram, interpreting a histogram

(a) Following are the ordered values of the property damages of the 50 states plus Puerto Rico. Use these values to answer the question by identifying the five states with the greatest and smallest tornado damages. The ordered values will also be helpful when counting the number of states in each class for part (b).

```
Damage ($Millions)
    0.00    0.05    0.09    0.10    0.24    0.26    0.27    0.34    0.53
    0.66    1.49    1.78    2.14    2.26    2.27    2.33    2.37    2.94
    3.47    3.57    3.68    4.42    4.62    5.52    7.42   10.64   14.69
   14.90   15.73   17.11   17.19   23.47   24.84   27.75   29.88   30.26
   31.33   37.32   40.96   43.62   44.36   49.28   49.51   51.68   51.88
   53.13   62.94   68.93   81.94   84.84   88.60
```

(b) When drawing a histogram:

1. Divide the range of values of the data into classes or intervals of equal length.

2. Count the number of data values that fall into each interval.

For this exercise the classes given are "$0 \leq$ damage < 10," "$10 \leq$ damage < 20," and so on. Complete the table of the number of states in each damage class. The count for the first class is done for you.

Average Tornado Damage ($Millions)

Class	Count
0 to <10	25
10 to <20	
20 to <30	
30 to <40	
40 to <50	
50 to <60	
60 to <70	
70 to <80	
80 to <90	
Total	

With this small number of data values it is both easy and instructive to draw the histogram by hand. To draw the histogram:

1. Mark the intervals on the horizontal axis and label the axis. Include the units.

2. Mark the scale for the counts or percents on the vertical axis. Label the axis.

3. Draw bars, centered over each interval, up to the height equal to the count or percent. There should be no space between the bars (unless the count for a class is zero, which creates a space between bars).

The vertical axis in a histogram can be either the count or the percent of the data in each interval. Changing the units from counts to percents will not affect the shape of the histogram. In the histogram on the next page, the first bar has been drawn for you. Complete the histogram using the information in the table in (b). Describe the shape center and spread of the distribution. Which states may be outliers?

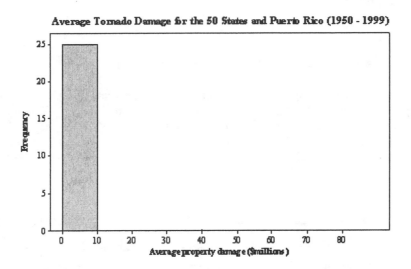

(c) Use your software to display the "default" histogram. How does it compare with your graph in (b)?

Exercise 1.41

KEY CONCEPTS: Drawing and interpreting a time plot, comparing two time plots

Complete the time plot on the graph. The number of problems per year for 1998 and 1999 for both GM and Toyota are included in the plot to get you started. When drawing two time plots on the same axes, it is helpful to connect the points for each time plot with a different type of line or color so that you can distinguish easily between the lines, particularly when they cross.

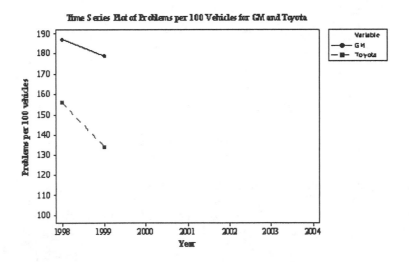

What is the general pattern in the time plot? What conclusions can you draw about trends over time and comparisons between the automakers?

COMPLETE SOLUTIONS

Exercise 1.1

(a) The individuals in this exercise are the different make and model cars.

(b) The variables are vehicle type (categorical), transmission type (categorical), number of cylinders (quantitative), city MPG (quantitative), and highway MPG (quantitative). Although the variable "cylinders" is quantitative, this variable divides the cars into only a few categories. To summarize this variable for a group of cars, we could draw a bar graph or a pie chart to give the percentage of 4-, 6-, or 8-cylinder cars in the group, even though these displays are typically used for categorical variables.

Exercise 1.3

(a) The percents in the table sum to 20 + 18 + 16 + 13 + 10 + 7 + 6 = 90. Thus the remaining 10% of the vehicles are some other color.

(b) The completed bar chart produced by MINITAB follows. If we added an "Other" category including the remaining 10% of the cars, it would be correct to make a pie chart as the categories would now make up the whole range of colors. Without the "Other" category, a pie chart would not be appropriate.

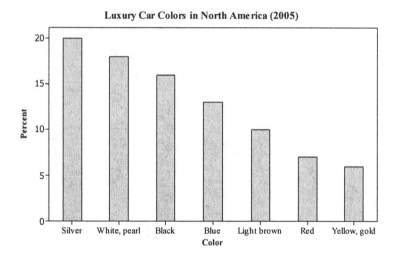

Exercise 1.11

Following is the completed stemplot. The distribution is skewed to the right with one clear outlier corresponding to 360 mg/dL. Only people in the second stem have been completely successful in maintaining glucose levels in the range 90 to 130 mg/dL, although the person at 80 does not have a problem with high levels. Most of the group still needs to work on lowering their glucose levels.

```
0 | 8
1 | 000134
1 | 5555677
2 | 0
2 | 67
3 | 6
```

Exercise 1.33

(a) The top five states for tornado damage are

```
Indiana          53.13
Illinois         62.94
Missouri         68.93
Oklahoma         81.94
Minnesota        84.84
Texas            88.60
```

and the bottom five are

```
Alaska            0.00
PuertoRico        0.05
RhodeIsland       0.09
Nevada            0.10
Vermont           0.24
```

(b) The distribution of average tornado damage over the period 1950 to 1999 for the 50 states and Puerto Rico and the corresponding histogram follow.

Average Tornado Damage ($Millions)

Class	Count
0 to <10	25
10 to <20	6
20 to <30	4
30 to <40	3
40 to <50	5
50 to <60	3
60 to <70	2
70 to <80	
80 to <90	3
Total	51

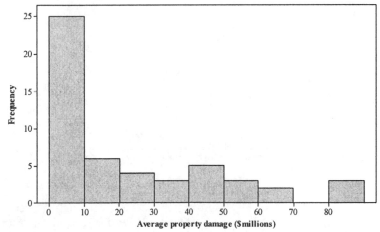

Average Tornado Damage for the 50 States and Puerto Rico (1950 - 1999)

The distribution is skewed to the right. Approximately half the states have damage below 10, while the remaining states have damage between 10 and 90. The center of the distribution is around 10, but because of the extreme skewness it is not a very good description of the "typical" damage value. The spread is from 0 to 88.6, although only the three states Oklahoma, Minnesota, and Texas have damage above 70. We would consider these three states outliers.

(c) The histogram below is the "default" histogram produced by MINITAB. It is identical to the one you have drawn by hand. However, the default choice of intervals selected by a computer package will not always be the ones you might choose.

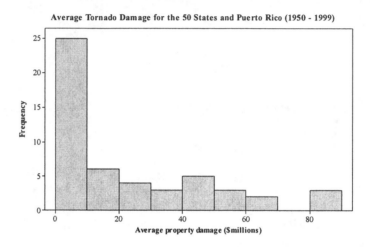

Exercise 1.41

(a) The two time plots are given in the following graph.

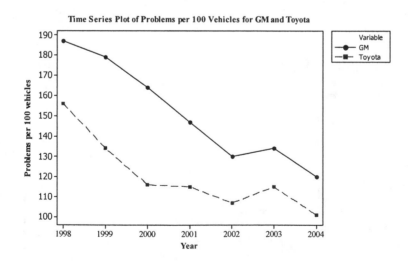

(b) Both companies have reasonably steady improvement in their car quality over this time period in terms of the number of "things gone wrong" in 90 days. However, on this measure, Toyota clearly outperforms GM over this time period.

CHAPTER 2

DESCRIBING DISTRIBUTIONS WITH NUMBERS

OVERVIEW

Once you have examined graphs to get an overall sense of the data, it is helpful to look at numerical summaries of features of the data that clarify the notions of center and spread.

Measures of center:
- **mean** (often written as $\bar{x}$)
- **median** (often written as M)

Finding the mean $\bar{x}$

The mean is the common arithmetic average. If there are n observations, x_1, x_2, L, x_n, then the mean is

$$\bar{x} = \frac{x_1 + x_2 + L + x_n}{n} = \frac{1}{n}\sum x_i$$

Recall that $\sum$ means "add up all these numbers."

Finding the median M

1. List all the observations from smallest to largest.

2. If the number of observations is odd, then the median is the middle observation. Count from the bottom of the list of ordered values up to the $(n + 1)/2$ largest observation. This observation is the median.

3. If the number of observations is even, then the median is the average of the two center observations.

Measures of spread:
- **quartiles** (often written as Q_1 and Q_3)
- **standard deviation** (s)
- **variance** (s^2)

Finding the quartiles Q_1 and Q_3

1.   Locate the median.

2.   The first quartile, Q_1, is the median of the lower half of the list of ordered observations.

3.   The third quartile, Q_3, is the median of the upper half of the list of ordered values.

Finding the variance s^2 and the standard deviation s

1.   Take the average of the squared deviations of each observation from the mean. In symbols, if we have n observations, x_1, x_2, L, x_n, with mean $\overline{x}$

$$s^2 = \frac{(x_1 - \overline{x})^2 + (x_2 - \overline{x})^2 + L + (x_n - \overline{x})^2}{n-1} = \frac{1}{n-1}\sum(x_i - \overline{x})^2$$

(Remember, $\sum$ means "add up.")

    If you are doing this calculation by hand, it is best to take it one step at a time. First calculate the deviations, then square them, next sum them up, and finally divide the result by $n - 1$. Exercise 2.9 follows the step-by-step calculations.

2.   The standard deviation is the square root of the variance: $s = \sqrt{s^2}$. Some things to remember about the standard deviation:

    (a) s measures the spread around the mean.

    (b) s should be used only with the mean, not with the median.

    (c) If $s = 0$, then all the observations must be equal.

    (d) The larger s is, the more spread out the data are.

    (e) s can be strongly influenced by outliers. It is best to use s and the mean only if the distribution is symmetric or nearly symmetric.

For measures of spread, the quartiles are appropriate when the median is used as a measure of center. In fact, the **five-number summary**, reporting the largest and smallest values of the data, the quartiles, and the median, provides a compact description of the data. The five-number summary can be represented graphically by a **boxplot**. If you use the mean as a measure of center, then the standard deviation and variance are the appropriate measures of spread. Watch out, because means and variances can be strongly affected by outliers and are harder to interpret for skewed data. The mean and standard deviation are not resistant measures. The median and quartiles are more appropriate when outliers are present or when the data are skewed. The median and quartiles are **resistant measures**.

GUIDED SOLUTIONS

Exercise 2.9

KEY CONCEPTS: Computing the mean, variance, and standard deviation

In practice, you will be using software or your calculator to obtain the mean and standard deviation from keyed-in data. This problem illustrates the step-by-step calculations to help you understand how the computations work. In general, be careful not to round off the numbers until the last step, as too-early rounding can introduce fairly large errors when computing s. However, in this example the numbers work out simply, so there should not be any rounding errors.

(a) Complete the step by step computation of the mean:

$$\bar{x} = \frac{1}{n}\sum x_i = \frac{x_1 + x_2 + \cdots + x_n}{n} =$$

(b) To do the step-by-step computation of the standard deviation, first complete the following table to obtain the sum of the squared deviations.

Observations	Deviations	Squared deviations
x_i	$x_i - \bar{x}$	$(x_i - \bar{x})^2$
5.6	0.2	0.04
	Sum =	Sum =

From the table we have $\sum (x_i - \bar{x})^2 =$

The variance is then computed as

$$s^2 = \frac{1}{n-1}\sum (x_i - \bar{x})^2 =$$

and the standard deviation as $s = \sqrt{s^2} =$

(c) Enter the data into your calculator and use the appropriate buttons to obtain $\bar{x}$ and s. Since no round off errors should have been introduced in your hand calculations, the results should agree.

Exercise 2.10

KEY CONCEPTS: Stemplots, mean and standard deviation, shape of distribution

Use statistical software or a calculator to find the mean and standard deviation of the two data sets and write in your answers here.

Data A $\bar{x} =$ $s =$
Data B $\bar{x} =$ $s =$

If you do the calculations correctly, you will find that these are two data sets of 11 observations with the same means and standard deviations. The mean gives an estimate of center, and the standard deviation gives an estimate of spread. Neither of these measures is resistant to outliers, and they do not give an indication of the shape of the distribution. Following is the stemplot of Data A, where the data have been rounded to the nearest 10th. The stems are 1s and the leaves are 10ths. Complete the stemplot of Data B and comment on the shapes of the two distributions.

Data A   Data B

```
3 | 1                                     5 |
4 | 7                                     6 |
5 |                                       7 |
6 | 1                                     8 |
7 | 3                                     9 |
8 | 1178                                 10 |
9 | 113                                  11 |
                                         12 |
```

Exercise 2.27

KEY CONCEPTS: Histograms, stemplots and boxplots, mean and standard deviation

If you are unclear on the details of computing the five-number summary, see Exercise 2.44, which goes through the step-by-step details. In this exercise we reproduce the five-number summaries for each variety and ask you to draw the boxplots for the *bihai* and red varieties by hand in the figure on the next page. The boxplot for the yellow variety is drawn for you.

Five number summaries:

	Bihai	Red	Yellow
Minimum	46.34	37.40	34.57
Q_1	46.71	38.07	35.45
M	47.12	39.16	36.11
Q_3	48.245	41.69	36.82
Maximum	50.26	43.09	38.13

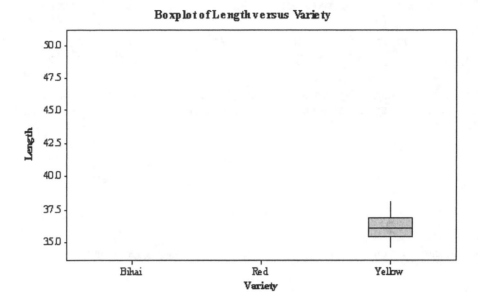

Do the boxplots fail to reveal any important information that is visible in the stemplots? What information is easily seen in both the stemplots and the boxplots?

Exercise 2.40

KEY CONCEPTS: Standard deviation

There are two points to remember in getting to the answer. The first is that numbers "further apart" from each other tend to have higher variability than numbers closer together. The other is that repeats are allowed. There are several choices for the answer to (a) but only one for (b).

Exercise 2.44

KEY CONCEPTS: Measures of center, five-number summary

(a) The five-number summary consists of the median, the minimum, and the maximum, and the first and third quartiles. To compute these quantities, it is simplest to first order the data. The ordered average property damages ($millions) are given on the next page. There are 51 observations as the data include all 50 states and Puerto Rico. What is the location of the median? Locate and underline the median in the list of ordered property damage on the next page.

0.00	0.05	0.09	0.10	0.24	0.26	0.27	0.34	0.53
0.66	1.49	1.78	2.14	2.26	2.27	2.33	2.37	2.94
3.47	3.57	3.68	4.42	4.62	5.52	7.42	10.64	14.69
14.90	15.73	17.11	17.19	23.47	24.84	27.75	29.88	30.26
31.33	37.32	40.96	43.62	44.36	49.28	49.51	51.68	51.88
53.13	62.94	68.93	81.94	84.84	88.60			

Remember that the first quartile is the median of the 25 observations below the median and that the third quartile is the median of the observations above the median. Underline the quartiles and fill in the following table.

Minimum

Q_1

M

Q_3

Maximum

Explain how you can see from these numbers that the distribution is right-skewed.

(b) Following is the histogram of the data.

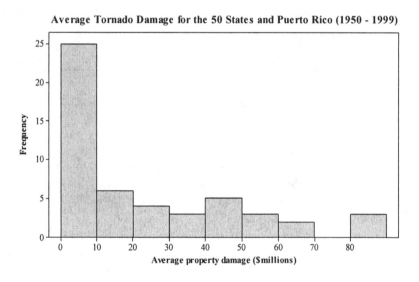

Average Tornado Damage for the 50 States and Puerto Rico (1950 - 1999)

The histogram suggests a few large outliers, namely the states above 80. To determine whether the 1.5 × IQR criterion flags the largest states, first compute the IQR. It is $Q_3 - Q_1$.

$IQR =$ $1.5 \times IQR =$

Then add $1.5 \times IQR$ to the third quartile and see if the damage from the largest state exceeds this value.

$$Q_3 + (1.5 \times IQR) =$$

You should find that there are no suspected outliers in the data according to this criterion.

(c) To find the mean, find the sum of the 51 observations and then divide by the number of observations.

Sum =

Mean =

Explain why the mean and median differ so greatly for this distribution.

COMPLETE SOLUTIONS

Exercise 2.9

(a) The step-by-step computation of the mean is

$$\bar{x} = \frac{1}{n}\sum x_i = \frac{x_1 + x_2 + \cdots + x_n}{n} = \frac{5.6 + 5.2 + 4.6 + 4.9 + 5.7 + 6.4}{5} = \frac{32.4}{5} = 5.4 \text{ mg of phosphate per}$$

deciliter of blood.

(b) To organize the calculations in the step-by-step computation of the standard deviation, we first complete this table to obtain the sum of the squared deviations.

Observations x_i	Deviations $x_i - \bar{x}$	Squared deviations $(x_i - \bar{x})^2$
5.6	0.2	0.04
5.2	-0.2	0.04
4.6	-0.8	0.64
4.9	-0.5	0.25
5.7	0.3	0.09
6.4	1.0	1.00
	Sum = 0.0	Sum = 2.06

From the table we have $\sum(x_i - \bar{x})^2 = 2.06$. The variance is computed as

$$s^2 = \frac{1}{n-1}\sum(x_i - \bar{x})^2 = \frac{2.06}{6-1} = 0.515,$$

and the standard deviation as $s = \sqrt{s^2} = \sqrt{0.412} = 0.6419$ mg of phosphate per deciliter of blood.

(c) The calculator gives the same results: $\bar{x} = 5.4$ and $s = 0.642$.

Exercise 2.10

Data A $\quad\quad \bar{x} = 7.501 \quad\quad\quad\quad s = 2.031$
Data B $\quad\quad \bar{x} = 7.501 \quad\quad\quad\quad s = 2.031$

From the following stemplots, we see two distributions with quite different shapes but with the same means and standard deviations. Data B seem less spread out than Data A despite the fact that the standard deviations are the same. The reason for this is that the standard deviation is not a resistant measure of spread and its value has been increased by the high outlier in Data B. Data A are left-skewed.

Data A   Data B

```
3 | 1                                    5 | 368
4 | 7                                    6 | 69
5 |                                      7 | 079
6 | 1                                    8 | 58
7 | 3                                    9 |
8 | 1178                               10 |
9 | 113                               11 |
                                      12 | 5
```

Exercise 2.27

The comparative boxplot follows.

Boxplot of Length versus Variety

The overall conclusions from the stemplots and boxplots are the same. The *bihai* variety is clearly the longest. All the *bihai* flower lengths are longer than any lengths of the other two varieties. The yellow variety is generally not as long as the red variety. The variability of the yellow and *bihai* varieties seem similar, and the red variety has much more variable lengths than either the yellow or *bihai*. An interesting feature that isn't apparent in the boxplot is the two possible outliers in the *bihai* group that appear in the stemplot. According to the $1.5 \times IQR$ rule there are *no* suspected outliers.

Exercise 2.40

(a) The standard deviation is always greater than or equal to zero. The only way it can equal zero is if all the numbers in the data set are the same. Since repeats are allowed, just choose four of the same numbers to make the standard deviation equal to zero. Examples are 1, 1, 1, 1 or 2, 2, 2, 2.

(b) To make the standard deviation large, numbers at the extremes should be selected, so you want to put the four numbers at 0 or 10. The correct answer is 0, 0, 10, 10. You might have thought 0, 0, 0, 10 or 0, 10, 10, 10 would be just as good, but a computation of the standard deviation of these choices shows that two at either end is the best choice.

(c) There are many choices for (a) but only one for (b).

Exercise 2.44

(a) The ordered observations are reproduced in the list that follows. Since there are 51 observations, the $(n + 1)/2$ largest observation is the $(51 + 2)/2 = 26$th largest observation. The median is in boldface and underlined. The first quartile is the median of the 25 observations below the median and is the $(n + 1)/2 = 13$th largest observation in this group. It is in italics and underlined. Similarly, the third quartile is the 13th largest observation among the 25 observations above the median, and it is also in italics and underlined.

0.00	0.05	0.09	0.10	0.24	0.26	0.27	0.34	0.53
0.66	1.49	1.78	*2.14*	2.26	2.27	2.33	2.37	2.94
3.47	3.57	3.68	4.42	4.62	5.52	7.42	**10.64**	14.69
14.90	15.73	17.11	17.19	23.47	24.84	27.75	29.88	30.26
31.33	37.32	*40.96*	43.62	44.36	49.28	49.51	51.68	51.88
53.13	62.94	68.93	81.94	84.84	88.60			

Here is the five-number summary.

Minimum	0.00
Q_1	2.14
M	10.64
Q_3	40.96
Maximum	88.60

You can see from the five-number summary that the smallest quarter of the observations are between 0 and 2 and the smaller half of the observations are between 0 and 10. The top half of the distribution is much more spread out, suggesting very strong skewness to the right.

(b) The *IQR* is $40.96 - 2.14 = 38.82$, and $1.5 \times IQR = 58.23$. If we add 58.23 to the third quartile, we get $40.96 + 58.23 = 99.19$. Since this exceeds the maximum, no states are flagged as suspected outliers.

(c) The sum of the 51 observations is 1119.60, the mean is $\bar{x} = 1119.60/51 = 21.95$ (1999 dollars), and the median is $M = 10.64$. The distribution is strongly right-skewed, and the mean exceeds the median because of this.

CHAPTER 3

THE NORMAL DISTRIBUTIONS

OVERVIEW

This chapter considers the use of **mathematical models** (mathematical formulas) to describe the overall pattern of a distribution. The name given to a mathematical model that summarizes the shape of a histogram is a **density curve**. The density curve is an idealized histogram. The area under a density curve between two x values represents the proportion of the data that lie between these two numbers. Like a histogram, a density curve can be described by measures of center such as the **median** (a point such that half the area under the density curve is to the left of the point) and the **mean** (the center of gravity or balance point of the density curve). In the last chapter we called the mean $\overline{x}$. The term refers to the mean of actual observations. The mean of a density curve is referred to as μ. Likewise, the standard deviation of a density curve also has a new notation. It is referred to as σ.

One of the most commonly used density curves in statistics is the **normal curve**. Normal curves are symmetric and bell-shaped, and as with all density curves the total area under the normal curve is 100%. Because the normal curve is symmetric around the mean, areas (proportions) such as those shown in the figure are equal.

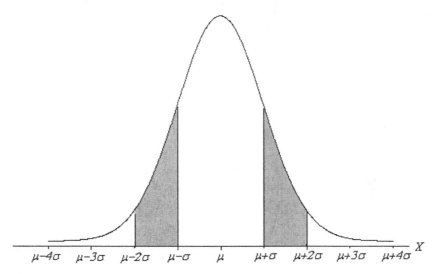

The peak of the normal curve is above the mean and the standard deviation measures the concentration of the area is around the peak.

Normal curves follow the 68–95–99.7 rule: 68% of the area under a normal curve lies within one standard deviation of the mean (illustrated in the figure on the next page), 95% within two standard deviations of the mean, and 99.7% within three standard deviations of the mean.

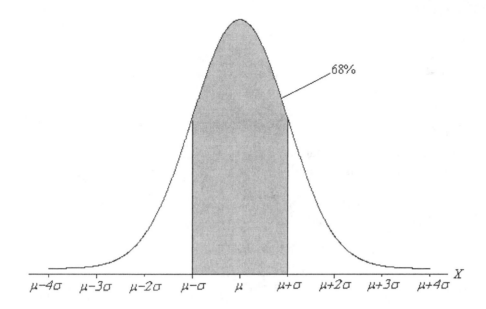

Areas under any normal curve can be found easily if an x value on the horizontal axis is first standardized by subtracting the mean (μ) from the x-value and dividing the result by the standard deviation (σ). This **standardized value** of x is called the **z-score**.

$$z = \frac{x - \mu}{\sigma}$$

If data whose distribution can be described by a normal curve are standardized (all values replaced by their z-scores), the distribution of the standardized values is described by the **standard normal curve**. Areas under standard normal curves are easily computed by using a **standard normal table**, such as Table A inside the front cover of your text.

The standard normal curve is very useful for finding the proportion of observations in an interval when dealing with any normal distribution. Here are some hints about solving these problems:

1. State the problem.

2. Draw a picture of the problem. It will help you know what area you are looking for.

3. Standardize the observations.

4. Using Table A inside the front cover of your text, find the area you need.

Use of statistical software: In the Guided Solutions, when looking in the body of Table A for a z-value corresponding to a given proportion, we use the value in the table closest to this proportion to solve the problem. We also give the answer that you would obtain if you used statistical software to get the z-value corresponding to this proportion.

GUIDED SOLUTIONS

Exercise 3.7

KEY CONCEPTS: 68–95–99.7 rule for normal density curves

(a) Recall the 68–95–99.7:

68% of the data will be between $\mu - \sigma$ and $\mu + \sigma$,
95% of the data will be between $\mu - 2\sigma$ and $\mu + 2\sigma$, and
99.7% of the data will be between $\mu - 3\sigma$ and $\mu + 3\sigma$.

What are μ and σ in this problem? Use the answer to find the values between which 99.7% of all pregnancies fall. The following figure should help you visualize the rule.

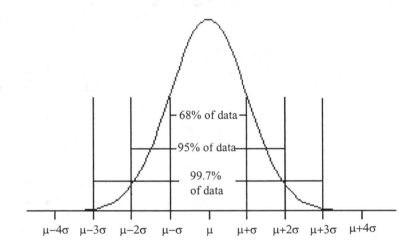

(b) Refer to the figure in (a). Below what value in the figure does 2.5% of the data fall? It is one of the values $\mu - 3\sigma$, $\mu - 2\sigma$, or $\mu - \sigma$. Remember that the normal curve is symmetric about μ.

Now convert this value to a value in days.

Exercise 3.9

KEY CONCEPTS: Standardized scores, z-scores

To compare scores from two normal distributions, each can be standardized or converted into a z-score. For example, a man's height or a woman's height that corresponds to a z-score greater than two places a man in the top 2.5% of men's heights or a woman in the top 2.5% of women's heights. This is because in either case the z-score corresponds to a height that is at least two standard deviations above the mean of its distribution. Using the mean and standard deviation from each distribution, convert a height of 6 feet (72 inches) to a z-score for men and women.

Women z-score =

Men z-score =

Which is more unusual: a 6-foot man or a 6-foot woman? The z-score helps answer this question.

Exercise 3.14

KEY CONCEPTS: Finding the value x (the quantile) corresponding to a given area under an arbitrary normal curve

This is an example of a "backward" normal calculation. First we *state the problem*. To make use of Table A on the inside front cover of your text, we need to state the problem in terms of areas to the left of some value. Next we *use the table*. To do so, we think of having standardized the problem and we then find the value z in the table for the standard normal distribution that satisfies the stated condition – that has the desired area to the left of it. We next must *unstandardize* this z-value by multiplying by the standard deviation and then adding the mean to the result. The unstandardized value x is the desired result. We illustrate this strategy in the solutions.

(a) *State the problem.* We are told that the Wechsler Adult Intelligence Scale scores are normally distributed with $\mu = 100$ and $\sigma = 15$. We want to find the score x that will place a score in the lowest 25% of this distribution. This means that 25% of the population scores are less than x. If the problem asked you to find the first quartile of the distribution, it would be another way of asking exactly the same question. We first need to find the corresponding value z for the standard normal. This is illustrated in the figure.

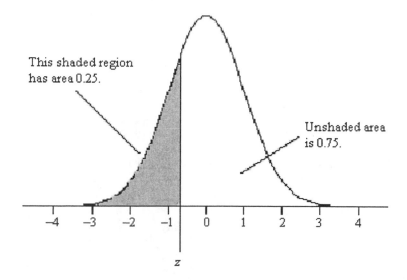

Use the table. The value z must have the property that the area to the left of it is 0.25. Areas to the left are the types of areas reported in Table A. Find the entry in the body of Table A that has a value closest to 0.25. This entry is 0.2514. The value of z that yields this area is seen from Table A to be -0.67.

Unstandardize. We now unstandardize z. The unstandardized value is

$$x = (\text{standard deviation}) \times z + \text{mean} = 15z + 100 = 15 \times (-0.67) + 100 = 89.95$$

Thus a person must score below 89.95 to be in the lowest 25%. Assuming that fractional scores are not possible, a person would have to score 89 or less to score below the first quartile.

Statistical software: Using statistical software to get the value of z corresponding to 0.25 instead of using the closest value in Table A gives $z = 0.6745$ and $x = 89.88$.

(b) Determine the score a person needs to place in the top 5%. Use the same line of reasoning as in (a).

State the problem. You may find it helpful to draw the region representing the z-value corresponding to the top 5%.

Use the table.

Unstandardize.

Exercise 3.30

KEY CONCEPTS: Computing areas under a standard normal density curve

Recall that the proportion of observations from a standard normal distribution that are less than a given value z is equal to the area under the standard normal curve to the left of z. Table A gives these areas. The areas are illustrated in the figure.

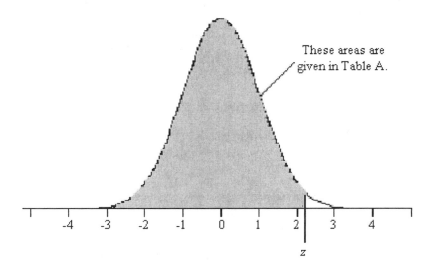

These areas are given in Table A.

In answering questions concerning the proportion of observations from a standard normal distribution that satisfy some relation, we find it helpful to first draw a picture of the area under a normal curve corresponding to the relation. We then try to visualize this area as a combination of areas of the form in the previous figure, since such areas can be found in Table A. The entries in Table A are then combined to give the area corresponding to the relation of interest.

This approach is illustrated in the solutions that follow.

(a) To get started, we will work through a complete solution. Following is a picture of the desired.

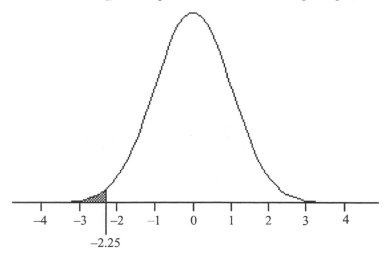

This is exactly the type of area that is given in Table A. We simply find the row labeled –2.2 along the left margin of the table, locate the column labeled .05 across the top of the table, and read the entry in the intersection of the row and column: 0.0122. This is the proportion of observations from a standard normal distribution that satisfies $z < -2.25$.

(b) Shade the desired area in the figure.

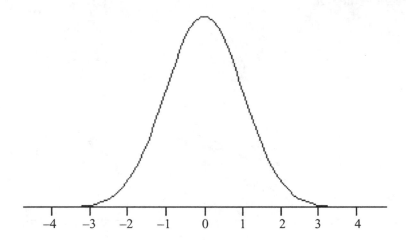

Remembering that the area under the whole curve is 1, how would you modify your answer from part (a)?

area =

(c) Try solving this part on your own. To begin, draw a picture of a normal curve and shade the region.

Now use the same line of reasoning as in part (b) to determine the area of your shaded region. Remember, you want to try to visualize the shaded region as a combination of areas of the form given in Table A.

(d) This part is a bit more complicated than the previous parts, but the same approach will work. Draw a picture and then try to express the desired area as the difference of two regions for which the areas can be found directly in Table A.

Exercise 3.33

KEY CONCEPTS: Standardized scores, z-scores

To compare scores from two normal distributions, each score can be standardized or converted into a z-score. For example, an ACT or an SAT score that corresponds to a z-score greater than two places either an ACT score in the top 2.5% of ACT scores or an SAT score in the top 2.5% of SAT scores. The reason is that in either case the z-score corresponds to a score that is at least two standard deviations above the mean of its distribution. Using the mean and standard deviation from each distribution, convert an ACT score of 16 and an ACT score of 670 to a z-score for each distribution.

ACT z-score =

SAT z-score =

Who has the higher score, Jacob or Emily? The z-score helps answer the question.

Exercise 3.41

KEY CONCEPTS: Finding the value x (the quantile) corresponding to a given area under an arbitrary normal curve

State the problem. We know that the ACT scores are roughly normally distributed with $\mu = 20.9$ and $\sigma = 4.8$. We want to find the score x that will place Abigail in the top 20% of the population of students. This means that 80% of the population scores are less than x. We will need to first find the corresponding value z for the standard normal distribution, illustrated in the figure on the next page.

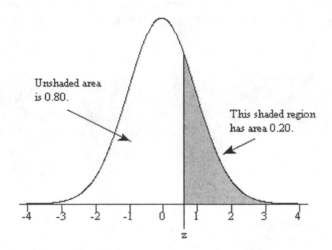

Use the table. The value *z* must have the property that the area to the left of it is 0.80. Areas to the left are the types of areas reported in Table A. Find the entry in the body of Table A that has a value closest to 0.80. Then find the value of *z* that yields this area.

Entry closest to 0.80 is

Value of *z* corresponding to the entry of 0.80 is

Unstandardize. We now unstandardize *z*. Using the value of *z* corresponding to the entry of 0.80 in Table A, the unstandardized value is

$$x = (\text{standard deviation}) \times z + \text{mean} = 4.8 \times z + 20.9 =$$

What score would Abigail need to achieve to be in the top 20%?

COMPLETE SOLUTIONS

Exercise 3.7

(a) In this problem, the mean is $\mu = 266$ days and the standard deviation is $\sigma = 16$ days. From the 68–95–99.7 rule we know that the middle 99.7% of all pregnancies should fall between

$$\mu - 3\sigma = 266 - (3 \times 16) = 266 - 48 = 218 \text{ days and}$$

$$\mu + 3\sigma = 266 + (3 \times 16) = 266 + 48 = 314 \text{ days}$$

(b) Since 95% of all pregnancies fall between $\mu - 2\sigma = 234$ days and $\mu + 2\sigma = 298$ days, the remaining 5% of all pregnancies should last less than 234 days or more than 298 days. The symmetry of the normal curve about its mean implies that half of this 5% (in other words, 2.5%) will be below 234 days and the remaining 2.5% above 298 days. Thus the shortest 2.5% of all pregnancies are less than 234 days.

Exercise 3.9

Women $\quad z\text{-score} = \dfrac{72 - 64}{2.7} = 2.96$

Men $\quad z\text{-score} = \dfrac{72 - 69.3}{2.8} = 0.96$

Rounding the z-scores of 2.96 and 0.96 to 3 and 1, respectively, and then applying the 68–95–99.7 rule, we see that about 0.15% of women are at least 6 feet tall while about 16% of men are at least 6 feet tall.

Exercise 3.14

(a) The complete solution for the lowest 25% is given in the Guided Solutions.

(b) For the top 5% we proceed as follows.

State the problem. We are told in this exercise that the Wechsler Adult Intelligence Scale scores are normally distributed with $\mu = 100$ and $\sigma = 15$. We want to find the score x that will place a person in the top 5% of this distribution. This means that 95% of the population scores are less than x. We first need to find the corresponding value z for the standard normal illustrated in the figure.

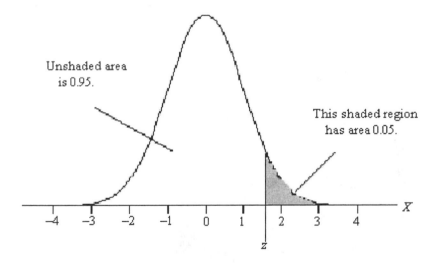

Use the table. The value z must have the property that the area to the left of it is 0.95. Areas to the left are the types of areas reported in Table A. Find the entry in the body of Table A that has a value closest to 0.95. This entry is 0.9495. The value of z that yields this area is seen from Table A to be 1.64.

Unstandardize. We now unstandardize z:

$$x = (\text{standard deviation}) \times z + \text{mean} = 15z + 100 = 15 \times 1.64 + 100 = 124.6$$

Thus a person must score at least 124.6 to be in the top 5%. Assuming that fractional scores are not possible, a person would have to score at least $x = 125$ to place in the top 5%. (The use of $z = 1.65$ gives $x = 124.75$)

Statistical software: Using statistical software to get the value of z corresponding to 0.95 instead of using the closest value in Table A gives $z = 1.6449$ and $x = 124.67$.

Exercise 3.30

(a) A complete solution was provided in the Guided Solutions.

(b) The desired area is indicated in the figure.

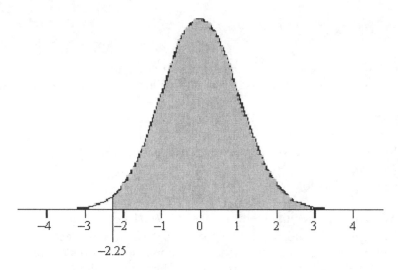

This is not of the form for which Table A can be used directly. However, the unshaded area to the left of –2.25 is of the form needed for Table A. In fact, we found the area of the unshaded portion in part (a). We notice that the shaded area can be visualized as what is left after deleting the unshaded area from the total area under the normal curve.

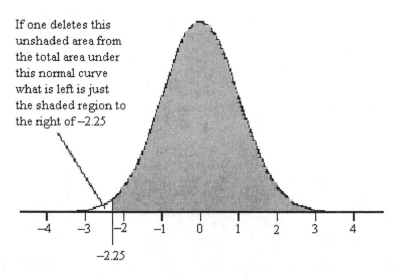

Since the total area under a normal curve is 1, we have

> shaded area = total area under normal curve – area of unshaded portion
>
> = 1 – 0.0122 = 0.9878.

Thus the desired proportion is 0.9878.

(c) The desired area is indicated in the figure.

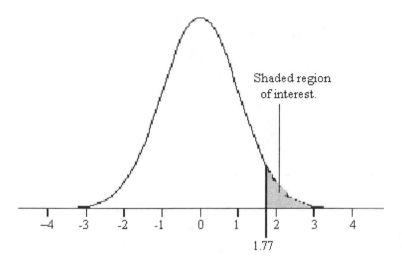

As in part (b) the unshaded area to the left of 1.77 can be found in Table A and is 0.9616. Thus

shaded area = total area under normal curve − area of unshaded portion

$$= 1 - 0.9616 = 0.0384$$

This is the desired proportion.

(d) We begin with a picture of the desired area.

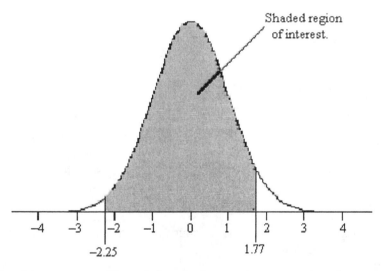

The shaded region is a bit more complicated than in the previous parts, but the same strategy works. We note that the shaded region is obtained by removing the area to the left of −2.25 from all the area to the left of 1.77.

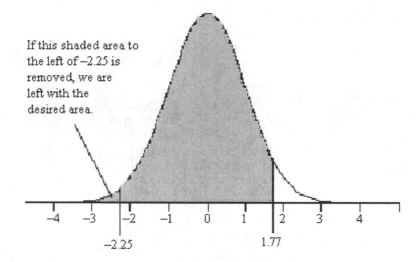

If this shaded area to the left of −2.25 is removed, we are left with the desired area.

−2.25

1.77

The area to the left of −2.25 is found in Table A to be 0.0122. The area to the left of 1.77 is found in Table A to be 0.9616. Thus

shaded area = area to left of 1.77 − area to left of −2.25

$$= 0.9616 - 0.0122$$

$$= 0.9494$$

This is the desired proportion.

Exercise 3.33

ACT $z\text{-score} = \dfrac{16 - 20.8}{4.8} = -1.00$

SAT $z\text{-score} = \dfrac{670 - 1026}{209} = -1.70$

Both Jacob and Emily have scored below the mean, putting them in the lower half of the scores. But Emily's SAT score of 670 is 1.7 standard deviations below the mean SAT score, while Jacob's ACT score is 1 standard deviation below the mean ACT score, so Jacob has the higher score.

Exercise 3.41

We find the entry in the body of Table A that has a value closest to 0.80: 0.7995. The value of z that yields this area is seen from Table A to be 0.84.

Unstandardize. The unstandardized value of z is

$$x = (\text{standard deviation}) \times z + \text{mean} = 4.8 \times z + 20.9 = 4.8 \times 0.84 + 20.9 = 24.93$$

Abigail must score at least 24.93 to be in the top 20%. Assuming that fractional scores are not possible, Abigail would have to score at least $x = 25$ to place in the top 20%.

Statistical software: Using statistical software to get the value of z corresponding to 0.80 instead of using the closest value in Table A gives $z = 0.8416$ and $x = 24.84$.

CHAPTER 4

SCATTERPLOTS AND CORRELATION

OVERVIEW

Chapters 1, 2, and 3 of your textbook provide tools for exploring several types of variables one by one. However, in most instances the data of interest are a collection of variables that may exhibit relationships among themselves. Typically, these relationships are more interesting than the behavior of the variables individually. In this chapter, we consider tools for exploring the relationship between variables. The first tool we consider is the **scatterplot**. Scatterplots display two quantitative variables at a time, such as the weight of a car and its miles per gallon (mpg). Using colors or different symbols, we can add information to the plot about a third variable that is categorical. For example, if in our plot we wanted to distinguish between cars with manual or automatic transmissions, we might use a circle to plot the cars with manual transmissions and a cross to plot the cars with automatic transmissions.

In a scatterplot, one variable is shown on the horizontal axis and the other is shown on the vertical axis. When there is a **response variable** and an **explanatory variable**, the explanatory variable is always placed on the horizontal axis. In cases where there is no explanatory-response variable distinction, either variable can go on the horizontal axis. After drawing the scatterplot by hand or using a computer, the scatterplot should be examined for an **overall pattern** that may tell us about any relationship between the variables and about **deviations** from it. You should be looking for the **direction, form,** and **strength** of the overall pattern. In terms of direction, **positive association** occurs when the variables both take on high values together, while **negative association** occurs if one variable takes on high values when the other takes on low values. In many cases, when an association is present, the variables appear to have a form that can be described as a **linear relationship**. The plotted values seem to form a line. If the line slopes up to the right, the association is positive; if the line slopes down to the right, the association is negative. As always, look for **outliers**. The outlier may be far away from the horizontal variable or the vertical variable or far away from the overall pattern of the relationship.

Categorical variables can be added to a scatterplot by using a different color or symbol for each category. If the response depends on the categorical variable, the overall pattern for points in different categories will differ, and differences will be evident in your plot.

Scatterplots provide a visual tool for looking at the relationship between two variables. Unfortunately, our eyes are not good tools for judging the strength of a relationship. Changes in the scale or the amount of white space in the graph can easily change our judgment of the strength of the relationship. **Correlation** is a numerical measure we use to show the strength of **linear association**.

The correlation can be calculated using the formula

$$r = \frac{1}{n-1} \sum \left(\frac{x_i - \bar{x}}{s_x}\right)\left(\frac{y_i - \bar{y}}{s_y}\right)$$

where $\bar{x}$ and $\bar{y}$ are the respective means for the two variables X and Y, and s_x and s_y are their respective standard deviations. In practice, you will probably be computing the value of r using computer software or a calculator that finds r from entering the values of the x's and y's. When computing a correlation coefficient, there is no need to distinguish between the explanatory and response variables, even in cases where this distinction exists. The value of r does not change if we switch x and y.

When r is positive there is a positive linear association between the variables, and when r is negative there is a negative linear association. The value of r is always between 1 and –1. Values close to 1 or –1 show a strong association, while values near 0 show a weak association. As with means and standard deviations, the value of r is strongly affected by outliers. Their presence can make the correlation much different than it might be with the outliers removed. Finally, remember that the correlation is a measure of straight-line association. There are many other types of association between two variables, but these patterns will not be captured by the correlation coefficient.

GUIDED SOLUTIONS

Exercise 4.1

KEY CONCEPTS: Explanatory and response variables

(a) When examining the relationship between two variables, if you hope to show that one variable can be used to explain variation in the other, remember that the response variable measures the outcome of the study, while the explanatory variable explains changes in the response variable. It is reasonable to assume in this example that the amount of time spent studying for a statistics exam can be used to explain the grade on the exam. Thus, we take the amount of time spent studying as the explanatory variable and the grade as the response variable.

(b) Should you explore the relationship between the variables or view one as explanatory and the other as a response?

If you believe you should view one as explanatory and the other as a response,

explanatory variable =   response variable =

(c) Should you explore the relationship between the variables or view one as explanatory and the other as a response?

If you believe you should view one as explanatory and the other as a response,

explanatory variable =   response variable =

(d) Should you explore the relationship between the variables or view one as explanatory and the other as a response?

If you believe you should view one as explanatory and the other as a response,

explanatory variable =   response variable =

Exercise 4.6

KEY CONCEPTS: Drawing and interpreting a scatterplot

(a) When drawing a scatterplot, one variable (the explanatory variable) is on the horizontal axis and the other (the response) is on the vertical axis. In this data set we are interested in the "effect" of speed on fuel used. So the speed that the car is driven is the explanatory variable and fuel used is the response variable, shown in the plot in the following figure. Although you will generally draw scatterplots on the computer, drawing a small one like this by hand makes sure that you understand what the points represent. The first two points correspond to speeds of 10 and 20 km/h. Complete the plot by hand.

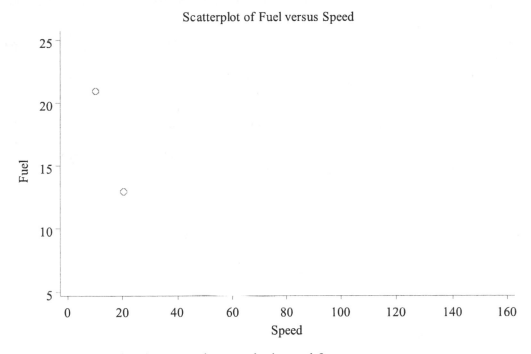

Scatterplot of Fuel versus Speed

(b) How would you describe the pattern in your plot in words?

Explain why this pattern makes sense. (In your experience, how does fuel consumption vary with speed?)

(c) Positive association occurs when the variables both take on high values together, while negative association occurs if one variable takes on high values when the other takes on low values. Is either of these patterns present here?

(d) How close to a simple curved pattern do the points in the plot lie? If the points lie close to a simple curve, we say that the relationship is strong. If the points are scattered around a curved pattern and do not lie close to it, we say that the relationship is weak.

Exercise 4.7

KEY CONCEPTS: Drawing and interpreting a scatterplot, adding a categorical variable to a scatterplot

(a) When drawing a scatterplot, place one variable (the explanatory variable) on the horizontal axis and the other (the response) on the vertical axis. In this data set we are interested in the "effect" of time on length. So time is the explanatory variable and length of the icicle is the response. To include the categorical variable, "run," use a different plotting symbol for the data from runs 8903 and 8905. Try using an o for points corresponding to run 8903 and an x for points corresponding to run 8905. The points corresponding to a time of 10 minutes and 20 minutes for both runs are drawn in the plot that follows. Complete the plot, either by hand or by using software.

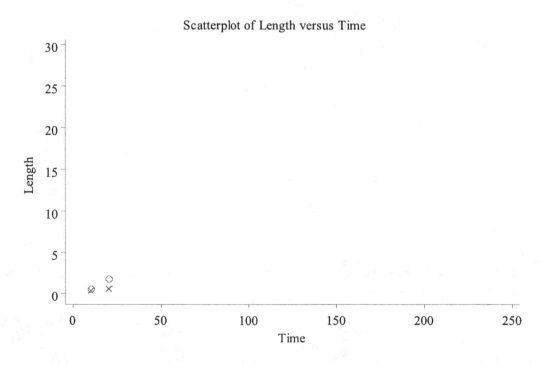

Scatterplot of Length versus Time

(b) Describe the form and strength of the relationship for the points corresponding to the different runs. Can the relationships be described with a straight line? How do the patterns for the two runs differ?

Exercise 4.8

KEY CONCEPTS: Scatterplots and computing the correlation coefficient

(a) For these data the statement of the question implies that prices explain forest lost. Higher coffee prices motivate farmers to clear more land for coffee trees. Thus the variable Price is the explanatory variable, and is plotted on the horizontal axis. Forest lost is the response, and is plotted on the vertical axis. Make your scatterplot on the axes provided.

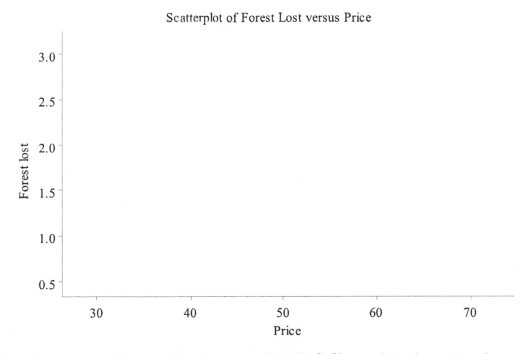

Scatterplot of Forest Lost versus Price

Do the points appear to lie approximately on a straight line? If not, what other pattern does your plot show?

(b) Let x denote the price and y the forest lost. We find the means and standard deviations to be

$$\bar{x} = 50.0 \qquad s_x = 16.32$$

$$\bar{y} = 1.738 \qquad s_y = 0.928$$

Calculations by hand are best done systematically, such as in the following table. The second and fourth columns are the standardized values for x and y. The table entries for the first two x, y values are provided. See if you can complete the remaining entries

x	$\left(\dfrac{x-\bar{x}}{s_x}\right)$	y	$\left(\dfrac{y-\bar{y}}{s_y}\right)$	$\left(\dfrac{x-\bar{x}}{s_x}\right)\left(\dfrac{y-\bar{y}}{s_y}\right)$
29	−1.2868	0.49	−1.3448	1.7305
40	−0.6127	1.59	−0.1595	0.0978
54		1.69		
65		1.82		
71		3.10		

Now sum up the values in the last column and divide by $n-1$ to compute r.

$r =$

(c) Use your calculator to compute r. You may need to consult your owner's manual if you do not know how to compute r using your calculator.

Exercise 4.32

KEY CONCEPTS: Interpreting the correlation coefficient

(a) The key sentence is "A well-diversified portfolio includes assets with low correlations." What does the sentence imply about which of the two investments she should choose?

(b) What can you say about the correlation between two variables when increases in one of the variables are associated with decreases in the other?

Exercise 4.33

KEY CONCEPTS: Changing units of measurement

(a) You can use software to construct the plot, or you can prepare it by hand on the axes provided.

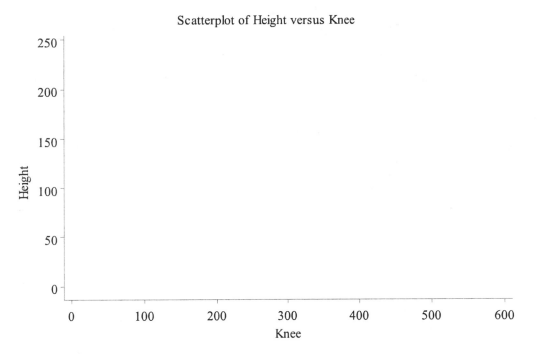

Scatterplot of Height versus Knee

(b) You may find it helpful to look at Fact 2 in the section on facts about correlation in your textbook. Now compute the two correlations.

Exercise 4.39

KEY CONCEPTS: Drawing and interpreting a scatterplot, computing and interpreting the correlation coefficient

The steps of the process are as follows:

State. What is the practical question in the context of the real-world setting?

Formulate. What specific statistical operations does the problem call for?

Solve. Make the graphs and carry out any calculations needed for the problem.

Conclude. Give your practical conclusion in the setting of the real-world problem.

To apply these steps to this problem, here are some suggestions.

State. The problem asks whether the data support the theory that a smaller percent of birds survive following a successful breeding season. What does the theory suggest about the relationship between breeding pairs and the percent that return the next season?

Formulate. What specific statistical operations do you need to do to explore this relationship?

Solve. Draw any plots and compute any quantities specified in the *Formulate* step.

Conclude. From the results of the *Solve* step, do the data support the theory that a smaller percent of birds survive following a successful breeding season? Why?

COMPLETE SOLUTIONS

Exercise 4.1

(a) A complete solution is provided in the Guided Solutions.

(b) We would probably simply want to explore the relationship between weight and height.

(c) We would probably view hours per week of extracurricular activities as explaining grade point average. The reasoning might be that the more time spent on extracurricular activities, the less time spent on study. Thus, the response is grade point average and the explanatory variable is hours per week of extracurricular activities.

(d) We would probably simply want to explore the relationship between a student's scores on the SAT math exam and student's SAT verbal exam.

Exercise 4.6

(a) The complete scatterplot follows.

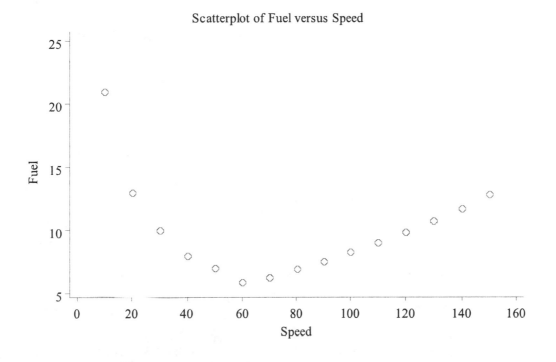

Scatterplot of Fuel versus Speed

(b) The plot shows a curved relationship. Fuel used first decreases as speed increases, and then at about 60 km/h increases as speed increases. This is not surprising. At very slow speeds and at very high speeds, engines are very inefficient and use more fuel, while at moderate speeds engines are more efficient and use less fuel.

(c) Variables are positively associated when both take on high values together and both take on low values together. Negative association occurs when high values of one variable are associated with low values of the other. In the scatterplot, both low and high speeds correspond to high values of fuel used, so we cannot say that the variables are positively or negatively associated.

(d) The points appear to lie close to a simple curve, so we would say the relationship is reasonably strong.

Exercise 4.7

(a) The scatterplot for all the data follows.

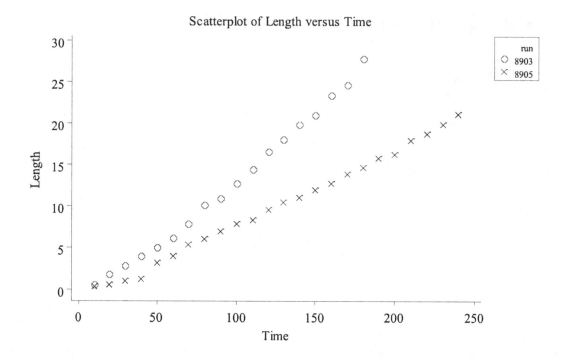

(b) For both runs, there is a strong straight line relationship between time and length. Not surprisingly, length increases with time. The rate of growth is clearly faster for run 8903 (the o's in the plot) that corresponds to the slower rate of water flow than for run 8905 (the x's in the plot).

Exercise 4.8

(a) The scatterplot follows.

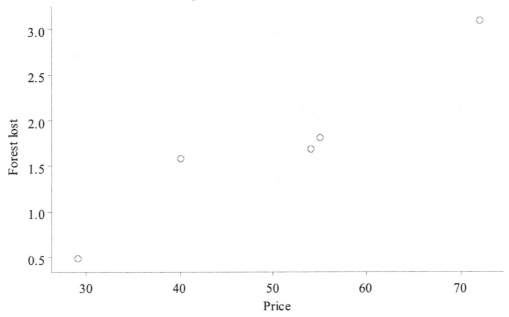

Scatterplot of Forest Lost versus Price

Although not perfect, the points do appear to lie approximately on a straight line.

(b) The completed table follows.

x	$\left(\dfrac{x-\overline{x}}{s_x}\right)$	y	$\left(\dfrac{y-\overline{y}}{s_y}\right)$	$\left(\dfrac{x-\overline{x}}{s_x}\right)\left(\dfrac{y-\overline{y}}{s_y}\right)$
29	−1.2868	0.49	−1.3448	1.7305
40	−0.6127	1.59	−0.1595	0.0978
54	0.2451	1.69	−0.0517	−0.0127
65	0.3064	1.82	0.0884	0.0271
71	1.3480	3.10	1.4677	1.9785

The sum of the values in the last column is 3.8212. Thus the correlation is

$$r = 3.8212/4 = 0.9553$$

(c) The calculator gives a correlation of 0.955, which agrees with the result in part (b).

Exercise 4.32

(a) Rachel should be looking for investments that have a weak correlation with municipal bonds, so she should invest in the small-cap stocks. The correlation between municipal bonds and the small-cap stock is weaker (closer to 0) than the correlation between municipal bonds and the large-cap stocks.

(b) When increases in one variable are associated with decreases in another, the two variables are negatively associated and the correlation is negative. Thus Rachel should look for assets that are negatively correlated with municipal bonds. Neither large-cap nor small-cap stocks meet this criterion.

Exercise 4.33

(a) The scatterplot follows. The points represented by the symbol o are the original data, and the points represented by the symbol x are those of the mad scientist. The patterns for the two sets of data look very different.

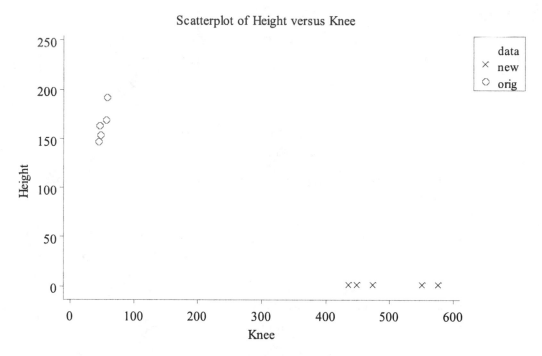

Scatterplot of Height versus Knee

(b) According to Fact 2 in the section on facts about correlation in your textbook, "Because *r* uses the standardized values of the observations, *r* does not change when we change the units of measurement of *x*, *y*, or both." Thus, we know without doing any calculations, that the correlation is exactly the same for the two sets of measurements. If you calculate the correlation for each, you find that both have a correlation of 0.877.

Exercise 4.39

State. A successful breeding season is indicated when the number of breeding pairs is high. The percent that return the next season is a measure of survival. According to the theory, if the number of breeding pairs is high (above average), the percent that return the next season will be low (below average). To see whether the data support the theory, we should examine the relationship between the number of breeding pairs and the percent that return the next breeding season.

Formulate. We should take the number of breeding pairs as the explanatory variable and the percent that return the next season as the response. We should make a scatterplot of the data to display the relationship between the variables and see if it supports the theory. If the relationship appears to be approximately linear, we should compute the correlation coefficient to quantify the strength of the relationship.

Solve. A scatterplot of the data follows.

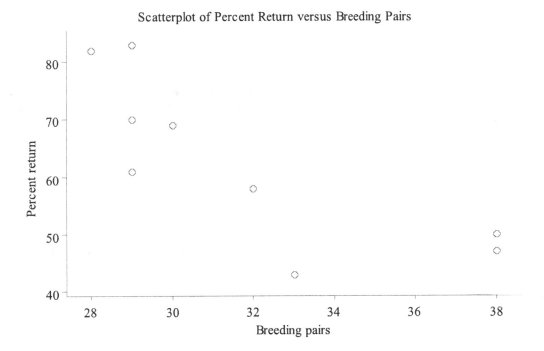

Scatterplot of Percent Return versus Breeding Pairs

The correlation coefficient is $r = -0.794$.

Conclude. The scatterplot shows a moderately strong negative association. The trend is roughly linear (although you may see a slight curve). The correlation coefficient is -0.794, confirming that the association is negative and that the relationship is moderately strong. The data support (or are consistent with) the theory that a smaller percent of birds survive following a successful breeding season.

CHAPTER 5

REGRESSION

OVERVIEW

If a scatterplot shows a linear relationship that is moderately strong as measured by the correlation, we can draw a line on the scatterplot to summarize the relationship. In the case where there is a response and an explanatory variable, the **least-squares regression** line often provides a good summary of the relationship. A straight line relating y to x has the form

$$y = a + bx$$

where b is the **slope** of the line and a is the **intercept**. The slope tells us the change in y corresponding to a one-unit increase in x. The intercept tells us the value of y when x is 0. This information has no practical meaning unless 0 is a feasible value for x.

The least-squares regression line is the straight line $\hat{y} = a + bx$, which minimizes the sum of the squares of the vertical distances between the line and the observed values y. The formula for the slope of the least-squares line is

$$b = r\frac{s_y}{s_x}$$

and for the intercept is $a = \bar{y} - b\bar{x}$, where $\bar{x}$ and $\bar{y}$ are the means of the x and y variables, s_x and s_y are their respective standard deviations, and r is the value of the correlation coefficient. Typically, the equation of the least-squares regression line is obtained by computer software or a calculator with a regression function. The least-squares regression line can be used to predict the value of y for any value of x. Just substitute the value of x into the equation of the least-squares regression line to get the predicted value for y.

Correlation and regression are clearly related, as can be seen from the equation for the slope b. However, the more important connection is how r^2 measures the strength of the regression. The square of the correlation coefficient, r^2, tells us the fraction of the variation in y that is explained by the regression of y on x. The closer r^2 is to 1, the better the regression describes the connection between x and y.

An examination of the **residuals** shows us how well our regression does in predictions. The difference between an observed value of y and the predicted value obtained by least-squares regression, $\hat{y}$, is called the residual.

$$residual = y - \hat{y}$$

Plotting the residuals is a good way to check the fit of a least-squares regression line. Features to look for in a **residual plot** are unusually large values of the residuals (outliers), nonlinear patterns, and uneven variation about the horizontal line through zero (corresponding to uneven variation about the regression line). Also look for influential observations. **Influential observations** are individual points whose removal would cause a substantial change in the regression line. Influential observations are often outliers in the horizontal direction.

Correlation and regression must be interpreted with caution.

• Do not **extrapolate.** To extrapolate means to predict beyond the range of the data.

• Be aware of possible **lurking variables**, variables that have an important effect on the relationship among the variables in a study but are not included among the variables studied.

• Are the data averages or from individuals? **Averaged data** usually lead to overestimating the correlations.

• Remember that *association is not causation!* Just because two variables are correlated doesn't mean that one causes changes in the other. The best evidence that an observed association is due to causation comes from a carefully designed experiment.

GUIDED SOLUTIONS

Exercise 5.4

KEY CONCEPTS: Drawing and interpreting the least-squares regression line, least-squares regression, prediction

(a) Use a calculator or statistical software to compute r. Write down your result.

$$r =$$

(b) The following results were obtained using the Minitab software package.

```
The regression equation is
new adults = 31.9 - 0.304 percent return


Predictor              Coef      SE Coef      T       P
Constant              31.934      4.838      6.60   0.000
percent return       -0.30402    0.08122    -3.74   0.003


S = 3.66689    R-Sq = 56.0%    R-Sq(adj) = 52.0%


Analysis of Variance

Source          DF      SS      MS      F       P
Regression       1    188.40  188.40  14.01   0.003
Residual Error  11    147.91   13.45
Total           12    336.31
```

The equation of the least-squares regression line is given at the beginning of the output. The intercept and slope can also be found in the table whose first column is labeled "Predictor." The column headed "Coef" gives the intercept (on the row labeled "Constant") and the slope (on the row labeled "percent return") in the same column. The values in this table are given to more decimal places than are in the equation at the beginning of the output.

From the output, we see that the equation of the least-squares regression line is

$$\text{new adults} = 31.9 - 0.304 \times (\text{percent return}).$$

To make a scatterplot of the data, use statistical software or the axes provided. Some software will also draw the least-squares regression line on the plot for you.

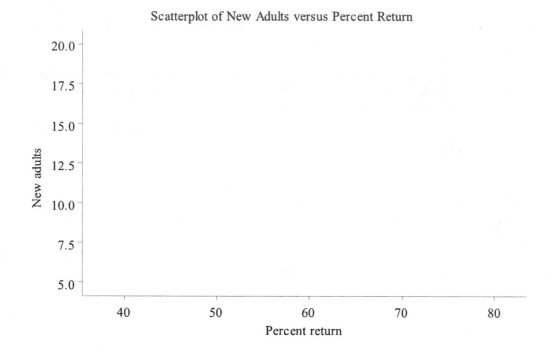

Scatterplot of New Adults versus Percent Return

Plot the regression line on this graph. Perhaps the simplest way to draw a graph of the least-squares regression line is to pick two convenient values for the variable Percent return, substitute them into the equation of the least-squares regression line, and compute the corresponding value of new adults predicted by the equation for each one. This process produces two sets of percent return and new adults values. Each of these percent return and new adults pairs corresponds to a point on the least-squares regression line. Simply plot these points on the graph and connect them with a straight line.

Convenient values for city mileage might be 40 and 80. Find the following values.

For percent return = 40, new adults =

For percent return = 80, new adults =

Plot these two sets of values on the graph. Then connect them with a straight line.

(c) The slope of the least-squares regression line is –0.304. Explain clearly in writing what this slope says about the change in the new adults. (Writing out your explanation forces you to express your thoughts explicitly. If you can't write a clear explanation, you may not adequately understand what the slope means.)

(d) To predict how many new birds will join another colony to which 60% of the adults from the previous year return, substitute percent return = 60 into the equation of the least-squares regression line.

new adults = $31.9 - 0.304 \times$ (percent return) =

Exercise 5.7

KEY CONCEPTS: Drawing a scatterplot, adding the least-squares regression line to the plot, prediction, residuals, plotting the residuals

(a) We discussed how to prepare a plot such as the one requested in this problem in the Guided Solution for Exercise 5.4. Use statistical software or the axes provided to create the plot.

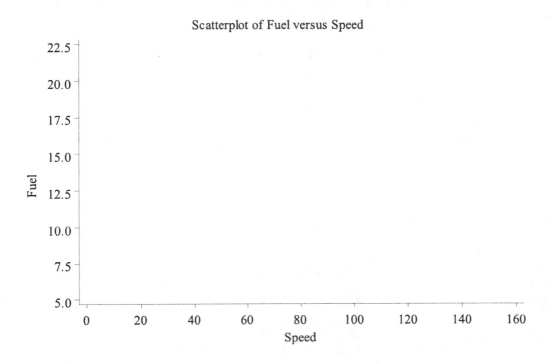

Scatterplot of Fuel versus Speed

48   Chapter 5

(b) What does your plot tell you about how well the least-squares regression line fits the data?

(c) We know that

residual = observed y – predicted y

For x (speed) = 10, what is the observed value of y (fuel) in the data?

$y =$

For $x = 10$, compute the predicted value of y using the least-squares regression line

$\hat{y} = 11.058 - 0.01466x =$

Now compute the residual.

residual =

Use a calculator or software to add up the values of the residuals given in the problem.

sum =

(d) Use statistical software or the axes provided to create a residual plot. Remember to add a horizontal line to the plot at height 0.

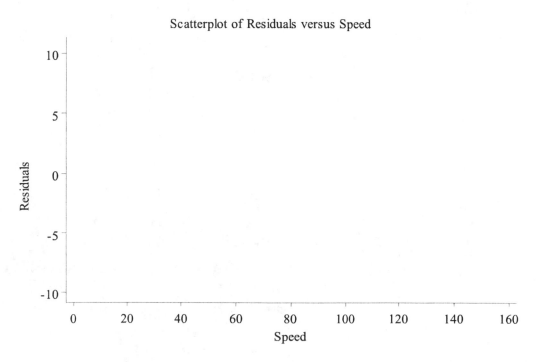

Scatterplot of Residuals versus Speed

How does the pattern of residuals about the horizontal line at height 0 compare to the pattern of the data points about the least-squares regression line in part (a)?

Exercise 5.8

KEY CONCEPTS: Outliers, influential observations

(a) Use statistical software or the axes provided to create the plot. Note that in the Guided Solution for Exercise 5.4, you made a scatterplot of the original 13 points. You may want to use that plot instead of drawing a new plot. Be sure to add points A and B to your plot. Use different plotting symbols to identify the points.

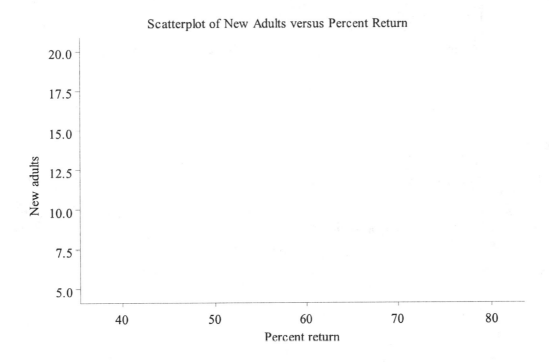

Scatterplot of New Adults versus Percent Return

In which direction is each new point an outlier?

Point A:

Point B:

(b)   Use a calculator or statistical software to find the least-squares regression line for each of the following.

• Original 13 points

  least-squares regression line:

• Original 13 points plus Point A

  least-squares regression line:

• Original 13 points plus Point B

  least-squares regression line:

Add these three lines to your scatterplot in part (a). See the Guided Solution to part (b) of Exercise 5.4 for suggestions on how to add the lines to your plot.

Which new point is more influential?

Why do the points move the line in the way your graph shows? Explain in simple language.

Exercise 5.9

KEY CONCEPTS: Scatterplots, least-squares regression, r^2, extrapolation

(a)   On the following axes, draw the scatterplot. Which is the response variable and which is the explanatory variable? Which goes on the vertical and which on the horizontal axis? Be sure to label the axes clearly.

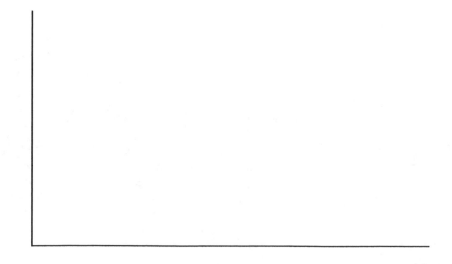

Use your calculator or statistical software to find the least-squares regression line. Write the equation.

(b) What quantity in the least-squares regression line represents the average decline per year in farm population during this period? What is its value? In answering this question, remember to keep in mind the units of the variable Population.

What quantity represents the percent of the observed variation in farm population that is accounted for by linear change over time? What is the value of this quantity?

(c) Use the equation of the least-squares regression line you computed in part (a) to answer this question. Again, remember the units of the variable Population in reporting your answer. Is your prediction reasonable? Why?

Exercise 5.31

KEY CONCEPTS: Calculating the least-squares regression line from summary statistics, r and r^2, prediction

(a) Recall that if the least-squares regression line has the equation

$$\hat{y} = a + bx$$

the formula for the slope of the least-squares line is

$$b = r\frac{s_y}{s_x}$$

and the formula for the intercept is

$$a = \bar{y} - b\bar{x}$$

where $\bar{x}$ and $\bar{y}$ are the means of the x and y variables, s_x and s_y are their respective standard deviations, and r is the value of the correlation coefficient.

What are the x and y variables in this problem?

Use the values for $\bar{x}$, $\bar{y}$, s_x, s_y, and r given in the problem to compute the slope b and intercept a.

$$b = \qquad\qquad a =$$

equation of least-squares regression line: $\hat{y} =$

(b) In part (a) you should have found that the equation of the least-squares regression line is

$$\text{final-exam score} = 30.2 + 0.16 \times (\text{pre-exam total})$$

Use this equation to predict the final-exam score for a student with a pre-exam total of 300 points.

predicted final-exam score =

(c) What quantity tells you the fraction of the variation in the values of y (in this case final-exam score) that is explained by the least-squares regression of y (or final-exam score) on x (in this case pre-exam total)? Compute this quantity. Remember to convert the fraction to a percent. What does this calculation tell you about the accuracy of the prediction?

Exercise 5.41

KEY CONCEPTS: Scatterplots, least-squares regression, correlation, r^2, the four-step process

The four-step process involves the following steps.

State. What is the practical question in the context of the real-world setting?

Formulate. What specific statistical operations does the problem call for?

Solve. Make the graphs and carry out any calculations needed for the problem.

Conclude. Give your practical conclusion in the setting of the real-world problem.

Here are some suggestions for applying the steps to this problem.

State. What does the problem ask you to do in the context of the real-world setting?

Formulate. Is there a graph that would help you uncover the *nature* of the effect of decreasing snow cover on wind stress?

What numerical quantities describe the *strength* of the effect of decreasing snow cover on wind stress?

Solve. Make graphs and calculate any quantities you identified in the formulate step.

Conclude. What do you conclude about the nature and strength of the effect of decreasing snow cover on wind stress?

Exercise 5.47

KEY CONCEPTS: Lurking variable

Think about what sort of person is likely to use artificial sweeteners. What is a lurking variable that prevents us from concluding that artificial sweeteners cause weight gain?

COMPLETE SOLUTIONS

Exercise 5.4

(a) Using software, we find the correlation to be

$$r = -0.748$$

(b) The equation of the least-squares regression line is given in the Guided Solution. A scatterplot with the regression line superimposed follows.

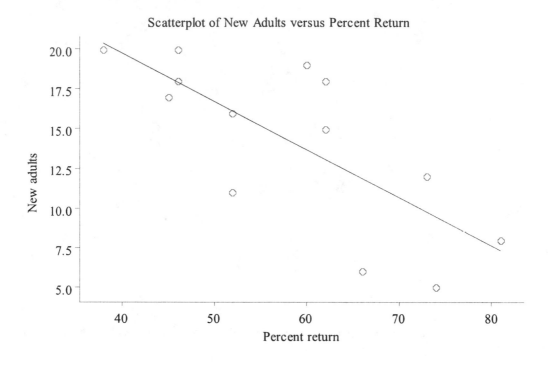

Scatterplot of New Adults versus Percent Return

(c) Slope = –0.304. This tells us that an increase of 1 percent in the number of adult sparrowhawks that return to a colony from the previous year on average corresponds to a decrease of 0.304 of new adult birds that join the colony.

(d) new adults = 31.9 – 0.304 × (percent return) = 31.9 – 0.304 × 60 = 13.66.

Exercise 5.7

(a) Using statistical software, we obtain the following plot.

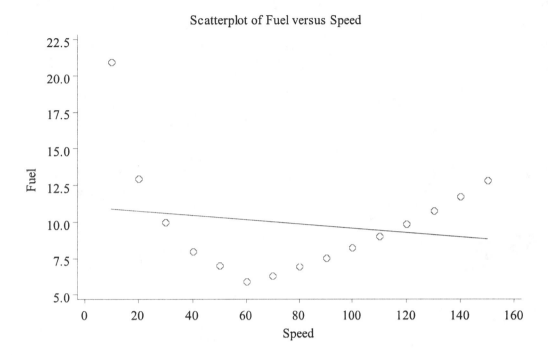

Scatterplot of Fuel versus Speed

(b) The data points clearly do not follow a straight line pattern. Thus, the least-squares regression line does not fit the data very well. As a consequence, I would not use the least-squares regression line to predict y from x.

(c) For x (speed) = 10, the observed value of y (fuel) in the data is

$$y = 21.00$$

For $x = 10$, compute the predicted value of y using the least-squares regression line

$$\hat{y} = 11.058 - 0.01466x = 11.058 - 0.01466 \times 10 = 10.91$$

Now compute the residual.

$$\text{residual} = \text{observed } y - \text{predicted } y = 21.00 - 10.91 = 10.09$$

Using a calculator, we find that

$$\text{sum of residuals} = -0.01$$

which differs from 0 only due to roundoff error.

(d) Using software, we obtain the plot on the next page. The pattern of points about the horizontal line at height 0 is almost identical to the pattern of the data points about the least-squares regression line in part (a). The only difference is that the pattern in this plot is tilted slightly compared to the pattern in the part (a) plot.

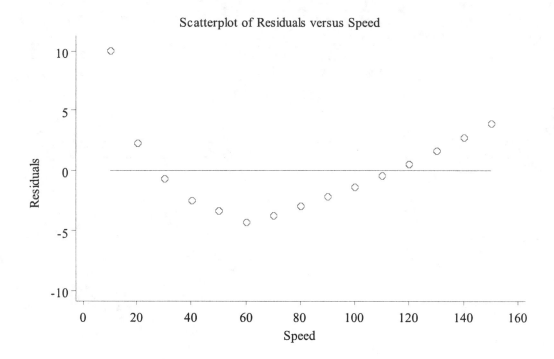

Scatterplot of Residuals versus Speed

Exercise 5.8

(a) Using statistical software, we obtain the following plot.

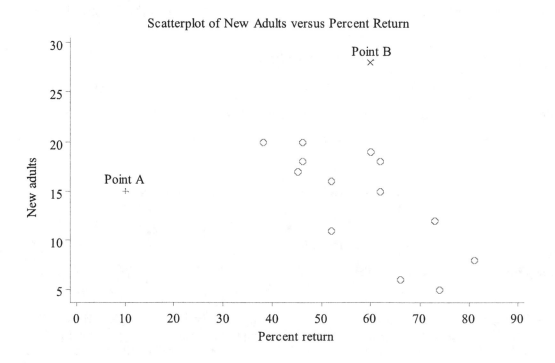

Scatterplot of New Adults versus Percent Return

We note that point A is an outlier in the *x* direction and point B is an outlier in the *y* direction.

(b) Using statistical software, we obtain the equations of the three least-squares regression lines.

• Original 13 points

  least-squares regression line: new adults = 31.9 - 0.304×(percent return)

• Original 13 points plus Point A

  least-squares regression line: new adults = 22.8 - 0.156×(percent return)

• Original 13 points plus Point B

  least-squares regression line: new adults = 32.3 - 0.293×(percent return)

We add the least-squares regression lines to the scatterplot.

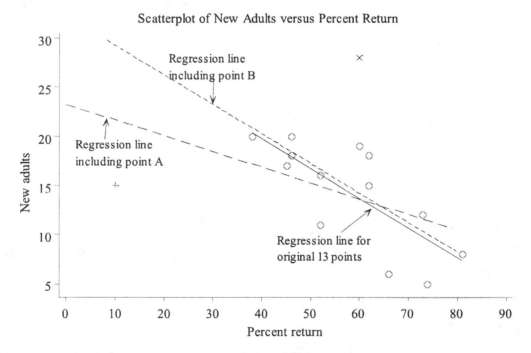

The plot clearly shows that point A is more influential than point B.

Point A is an outlier in the *x* direction. It lies well to the left of the original 13 points. Visually it has the effect of making the overall pattern in the plot look flatter. In other words, the overall pattern is best described by a "flatter" line.

Point B is an outlier in the *y* direction. It lies well above the original 13 points. Visually it has the effect of pulling the best fitting line up. In other words, it has little effect on the slope but increases the intercept slightly.

Exercise 5.9

(a) Here is a scatterplot of the data with year the explanatory variable and population the response.

Scatterplot of Population versus Year

We find that the least-squares regression line is

$$\text{population} = 1166.93 - 0.58679 \times \text{year}$$

(b) The slope of the least-squares regression line indicates a decline in farm population per year over the period represented by the data of 0.58679 million people per year, or 586,790 people per year. The percent of the observed variation in farm population accounted for by linear change over time is determined by the value of r^2, which is 0.977 (calculated using software). The desired percent is therefore 97.7%.

(c) In 2000 the regression equation predicts the number of people living on farms to be about

$$\hat{y} = 1166.93 - 0.58679(2000) = -6.65 \text{ million.}$$

This result is unreasonable since population cannot be negative. This prediction is an example of the dangers of extrapolation!

Exercise 5.31

(a) In the problem, the x variable is pre-exam total and the y variable is final-exam score. Given that

$$\bar{x} = 280 \qquad s_x = 30$$

$$\bar{y} = 75 \qquad s_y = 8$$

$$r = 0.6$$

the slope is

$$b = r\frac{s_y}{s_x} = 0.6\frac{8}{30} = 0.16$$

and the intercept is

$$a = \overline{y} - b\,\overline{x} = 75 - 0.16 \times 280 = 75 - 44.8 = 30.2$$

The equation of the least-squares regression line is therefore

$$\text{final-exam score} = 30.2 + 0.16 \times (\text{pre-exam total})$$

(b) We use the equation of the least-squares regression line we found in part (a), namely,

$$\text{final-exam score} = 30.2 + 0.16 \times (\text{pre-exam total})$$

to make our prediction. We calculate (Julie's pre-exam total is 300)

$$\text{predicted final-exam score} = 30.2 + 0.16 \times (300) = 30.2 + 48 = 78.2$$

(c) Recall that the square of the correlation, r^2, is the fraction of the variation in the values of y (in this case final-exam score) that is explained by the least-squares regression of y on x (in this case pre-exam total). Since here $r = 0.6$,

$$r^2 = (0.6)^2 = 0.36$$

Converting this fraction to a percent we find that the observed variation in these students' final-exam scores that can be explained by the linear relationship between final-exam score and pre-exam total is 36%. Thus, only a modest percentage of the observed variation in these students' final-exam scores can be explained by the linear relationship between final-exam score and pre-exam total. In other words, the prediction of Julie's final-exam score based on the equation of the least-squares regression line is probably not very accurate and her score could have been much higher (or much lower) than the predicted value of 78.2.

Exercise 5.41

State. The problem asks us to uncover the nature and strength of the effect of decreasing snow cover in the vast landmass of Europe and Asia on summer wind stress. Presumably higher values of summer wind stress imply stronger monsoons.

Formulate. A scatterplot of wind stress (the response) versus snow cover (the explanatory variable) provides information about the nature of the relationship between these variables. The least-squares regression line also provides information about the nature of this relationship. For example, the slope tells us whether decreasing snow cover is associated with increases or decreases in wind stress. Thus, it will be helpful to make both a scatterplot and compute the equation of the least-squares regression line.

The correlation r, or better yet, r^2, provides information about the strength of the relationship. Thus, it will be helpful to calculate both of these quantities.

Solve. Here is a scatterplot of wind stress versus snow cover.

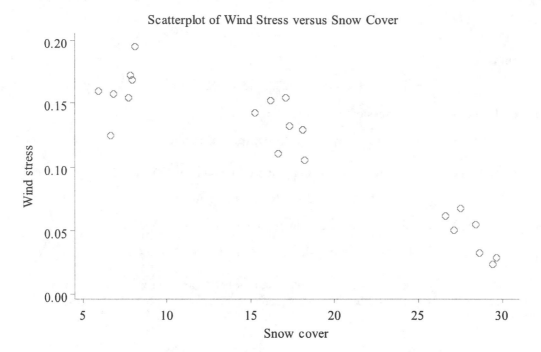

The plot shows an approximately linear trend with wind stress decreasing as snow cover increases.

The equation of the least-squares line computed using statistical software follows.

```
wind stress = 0.212 - 0.00561 × (snow cover)
```

The slope is negative, which is consistent with the trend observed in the scatterplot.

The values of r and r^2 are

$$r = 0.918, r^2 = 0.843$$

These values are reasonably large. In particular, the value of r^2 tells us that 84.3% of the variation in wind stress in the data is explained by the least-squares regression of wind stress on snow cover. We would describe the relation as strong.

Conclude. One must be careful about concluding cause and effect from such a study because it is not a designed experiment. However, the data suggest that decreasing snow cover is associated with greater wind stress. Furthermore, the relationship appears to be fairly strong. The data are certainly consistent with the hypothesis that decreasing snow cover causes an increase in wind stress.

Exercise 5.47

People who are overweight and trying to lose weight are among those likely to use artificial sweeteners. Thus, the weight of an individual when the person begins to use an artificial sweetener is a lurking variable. We would expect a larger proportion of overweight individuals to be among those using artificial sweeteners than among those using sugar, not because artificial sweeteners cause you to gain weight, but because overweight people are more likely to choose to use artificial sweeteners.

CHAPTER 6

TWO-WAY TABLES

OVERVIEW

This chapter discusses techniques for describing the relationship between two or more categorical variables. To analyze categorical variables, we use counts (frequencies) or percents (relative frequencies) of individuals that fall into various categories. **A two-way table** of such counts is used to organize data about two categorical variables. Values of the **row variable** label the rows that run across the table, and values of the **column variable** label the columns that run down the table. In each cell (intersection of a row and column) of the table, we enter the number of cases for which the row and column variables have the values (categories) corresponding to that cell.

The **row totals** and **column totals** in a **two-way table** give the marginal distributions of the two variables separately. It is usually clearest to present these distributions as percents of the table total. **Marginal distributions** do not give any information about the relationship between the variables. **Bar graphs** are a useful way of presenting these marginal distributions.

The **conditional distributions** in a two-way table help us see relationships between two categorical variables. To find the conditional distribution of the row variable for a specific value of the column variable, look only at that one column in the table. Express each entry in the column as a percent of the column total. There is a conditional distribution of the row variable for each column in the table. Comparing these conditional distributions is one way to describe the association between the row and column variables, particularly if the column variable is the explanatory variable. When the row variable is explanatory, find the conditional distribution of the column variable for each row and compare these distributions. Side-by-side bar graphs of the conditional distributions of the row or column variable can be used to compare these distributions and describe any association that may be present.

Data on three categorical variables can be presented as separate two-way tables for each value of the third variable. An association between two variables that holds for each level of this third variable can be changed, even reversed, when the data are combined by summing over all values of the third variable. **Simpson's paradox** refers to such reversals of an association.

GUIDED SOLUTIONS

Exercise 6.1

KEY CONCEPTS: Marginal distribution

(a) Each of the six entries in the table correspond to a different group of people. The six different groups characterize all the people that these data describe. Add up the six entries to determine how many people these data describe.

(b) The three entries in the row labeled "Arthritis" correspond to the people with arthritis of the hip or knee. Add up these three entries to get the total requested.

(c) First compute the counts for each level of participation in soccer (the columns) by adding the two entries in each column. Enter the totals in the following table in the row labeled "Counts." The marginal distribution, as a percent, can be found from the counts in each group (elite, non-elite, did not play) that you entered in the table. Each count must be divided by the total number of persons represented by the table that you computed in part (a). Now do the actual calculations and complete the table.

Group

	Elite	Non-elite	Did not play
Counts			
Percent			

Exercise 6.5

KEY CONCEPTS: Conditional distribution, describing relationships, the four-step process

The four-step process involves the following steps.

State. What is the practical question in the context of the real-world setting?

Formulate. What specific statistical operations does this problem call for?

Solve. Make the graphs and carry out any calculations needed for this problem.

Conclude. Give your practical conclusion in the setting of the real-world problem.

Here are some suggestions for applying the steps to this problem.

State. What issue regarding soccer and arthritis are you to investigate?

Formulate. What conditional distribution do you need to calculate to address the issue you identified in the *State* step?

Solve. Compute the conditional distributions and describe any patterns you see. To compute the conditional distribution of people with arthritis, first compute the total number of people in each group. To do this, add up the two entries in each column of the table in Exercise 6.1.

Total elite =

Total non-elite =

Total did not play =

To compute the percent of each group who have arthritis, divide each entry in the row labeled "Arthritis" by the total number in the group and convert to a percent. We have done the calculation for the elite players. Fill in the rest of the table to complete the exercise.

	Group		
	Elite	Non-elite	Did not play
Percent with arthritis	$10/71 \times 100\% = 14.08\%$		

What pattern do you see?

Conclude. What do you conclude about the relationship between serious soccer players and arthritis later in life?

Exercise 6.25

KEY CONCEPTS: Marginal and conditional distributions, describing relationships

(a) To compare the effectiveness of the three treatments in preventing relapse, compute the conditional distributions of relapse/no relapse for the three drugs. We have computed the percents for desipramine in the following table. Complete the table by computing the percents for lithium and placebo.

	Desipramine	Lithium	Placebo
Relapse	$(10/24) \times 100\% = 41.67\%$		
No relapse	$(14/24) \times 100\% = 58.33\%$		

After completing the table, plot the percentages in a bar graph. Use the following axes for your plot. For each of Relapse and No relapse, plot three bars.

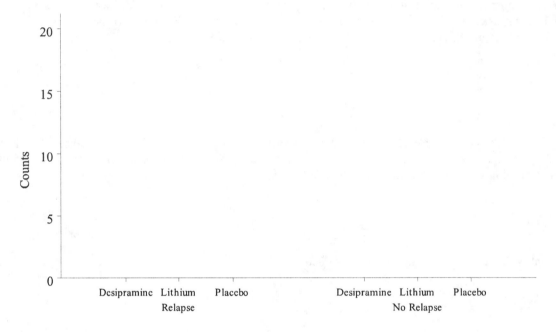

What do you observe?

(b) What types of studies provide the strongest evidence for causation? What things must one be concerned about when deciding whether an observed association is due to cause and effect?

Exercise 6.31

KEY CONCEPTS: Two-way tables, Simpson's paradox

(a) Add corresponding entries in the two tables and enter the sums in the table.

	Admit	Deny
Male		
Female		

(b) Convert your part (a) table to one involving percentages of the row totals.

	Admit	Deny
Male		
Female		

(c) Repeat the type of calculations you did in part (b) for each of the original tables.

Business

	Admit	Deny
Male		
Female		

Law

	Admit	Deny
Male		
Female		

(d) To explain the apparent contradiction observed in part (c), consider which professional school is easier to get into and which professional school males and females tend to apply to. Write your answer in plain English in the space provided. Avoid jargon, and be clear!

COMPLETE SOLUTIONS

Exercise 6.1

(a) The number of people these data describe is $10 + 9 + 24 + 61 + 206 + 548 = 858$.

(b) The number of people with arthritis of the hip or knee is $10 + 9 + 24 = 43$.

(c) Following is the completed table.

	Group		
	Elite	Non-elite	Did not play
Counts	10 + 61 = 71	9 + 206 = 215	24 + 548 = 572
Percent	71/858 × 100% = 8.28%	215/858 × 100% = 25.06%	572/858 × 100% = 66.67%

Exercise 6.5

State. We suspect that more serious soccer players have more arthritis later in life.

Formulate. We should calculate the conditional distribution of the different levels of experience playing soccer given arthritis later in life.

Solve.   Total elite = 10 + 61 = 71

Total non-elite = 9 + 206 = 215

Total did not play = 24 + 548 = 572

The percent of each group who have arthritis is given in the following table.

	Group		
	Elite	Non-elite	Did not play
Percent	10/71 × 100% = 14.08%	9/215 × 100% = 4.19%	24/572 × 100% = 4.20%

The percentage of cases of arthritis is higher (more than three times as high) for elite players than for non-elite players and those who did not play soccer. Non-elite players and those who did not play have nearly the same percentage of cases of arthritis.

Conclude. The data suggest that playing lots of soccer (as represented by elite players) is associated with later arthritis. We conclude that the data support the hypothesis that more serious soccer players have more arthritis later in life.

Exercise 6.25

(a) The conditional distributions of relapse/no relapse for the three drugs follows.

	Desipramine	Lithium	Placebo
Relapse	(10/24) × 100% = 41.67%	(18/24) × 100% = 75%	(20/24) × 100% = 83.33%
No relapse	(14/24) × 100% = 58.33%	(6/24) × 100% = 25%	(4/24) × 100% = 16.67%

Here is a bar graph that displays this information.s

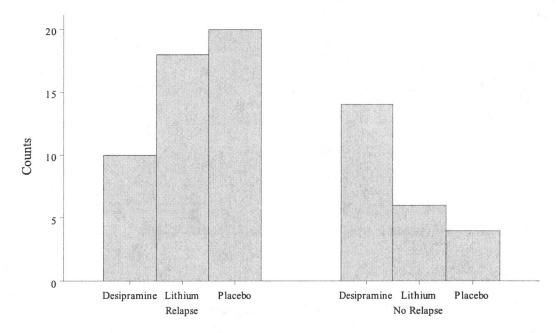

The data show that people taking desipramine had fewer relapses than people taking either lithium or a placebo.

(b) These results are interesting, but association does not by itself imply causation. Following are some concerns. First, how strong is the association? Is the number of people in the study large enough that we can conclude that the observed association is strong? Second, can we rule out lurking variables? Because this was an experiment in which the explanatory variable was directly changed and other influences on the response were controlled (by assigning subjects to treatments at random and using a placebo), the case for causation is strengthened. However, did the subjects know anything about the treatments they received and would this knowledge affect their response to the treatment? Third, have similar results been observed in other studies with other groups in other contexts? Without additional information, we should be cautious about concluding that the study demonstrates that desipramine causes a reduction in relapses.

Exercise 6.31

(a) Here is the desired two-way table.

	Admit	Deny
Male	490	210
Female	280	220

(b) We first add a column containing the row totals to the table in part (a).

	Admit	Deny	Total
Male	490	210	700
Female	280	220	500

We now convert the table entries to percents of the row totals. We divide the entries in the first row by 700 and express the results as percents. We divide the entries in the second row by 500 and express them as percents.

	Admit	Deny
Male	70%	30%
Female	56%	44%

We see that Wabash admits a higher percent of male applicants.

(c) We repeat the calculations in part (b), for each of the original two tables.

Business

	Admit	Deny	Total
Male	480	120	600
Female	180	20	200

Law

	Admit	Deny	Total
Male	10	90	100
Female	100	200	300

Converting entries to percents of the row totals yields

Business

	Admit	Deny
Male	80%	20%
Female	90%	10%

	Admit	Deny
Male	10%	90%
Female	33.3%	66.7%

We see that each school admits a higher percentage of female applicants.

(d) Although both schools admit a higher percentage of female applicants, the admission rates are quite different. Business admits a high percentage of all applicants; it is easier to get into the business school. Law admits a lower percentage of applicants; it is harder to get into the law school. Most of the male applicants to Wabash apply to the business school with its easy admission standards. Thus, overall, a high percentage of males are admitted to Wabash. The majority of female applicants apply to the law school. Because it has tougher admission standards, this makes the overall admission rate of females appear low, even though more females are admitted to both schools!

CHAPTER 7

EXPLORING DATA: PART I REVIEW

To assist you in reviewing the material in Chapters 1 – 6, we provide the text chapter and related exercises in this Study Guide for each of the odd-numbered review exercises. Other than pointing you in the right direction, we provide no additional hints or solutions. At this point, you should be able to work these exercises on your own with minimal assistance. As a final challenge, we encourage you to work some of the Supplementary Exercises, which integrate more fully the material in these chapters.

Exercise 7.1
Text location – Chapter 1 for types of variables
Related Study Guide exercise – Exercise 1.1

Exercise 7.3
Text location – Chapter 1 for time plots
Related Study Guide exercise – Exercise 1.41

Exercise 7.5
(a) Text location – Chapter 4 for measuring linear association: correlation
Related Study Guide exercise – Exercise 4.8

(b) Text location – Chapter 5 for the least-squares regression line
Related Study Guide exercise – Exercise 5.4

(c) Text location – Chapter 5 for the least-squares regression line and cautions about correlation and regression
Related Study Guide exercises – Exercises 5.4, 5.9

Exercise 7.7
Text location – Chapter 2 for means and standard deviations, Chapter 3 for 68–95–99.7 rule
Related Study Guide exercises – Exercises 2.9, 2.10, 3.7

Exercise 7.9
Text location – Chapter 2 for five-number summary and boxplots
Related Study Guide exercise – Exercise 2.39 is one method.

Exercise 7.11
(a) Text location – Chapter 1 for stemplots
Related Study Guide exercise – Exercise 1.11

(b) Text location – Chapter 2 for means and standard deviations
Related Study Guide exercises – Exercises 2.9, 2.10

Exercise 7.13
Text location – Chapter 2 for spotting suspected outliers
Related Study Guide exercise – Exercise 2.44

Exercise 7.15
Text location – Chapter 1 for drawing bar graphs
Related Study Guide exercise – Exercise 1.3

Exercise 7.17
(a) Text location – Chapter 1 for drawing histograms.
Related Study Guide exercise – Exercise 1.33

(b) Text location –Chapter 2 for mean, median, standard deviation, and five-number summary.
Related Study Guide exercises – Exercises2.9, 2.44

(c) Text location –Chapter 2 for mean, median, standard deviation, and five-number summary.
Related Study Guide exercises – Exercises 2.9, 2.44

Exercise 7.19
Text location – Chapter 1 for drawing histograms or stemplots, Chapter 2 for numerical summaries
Related Study Guide exercises – Exercises 1.11, 1.33, 2.9, 2.44

Exercise 7.21

(a) Text location – Chapter 4 for displaying relationships: scatterplots
Related Study Guide exercise – Exercise 4.6

(b) Text location – Chapter 4 for interpreting scatterplots
Related Study Guide exercise – Exercise 4.6

(c) Text location – Chapter 4 for interpreting scatterplots
Related Study Guide exercise – Exercise 4.6

Exercise 7.23

(a), (b), (c) Text location – Chapter 2 for numerical summaries
Related Study Guide exercise – Exercise 2.9

(d) Text location – Chapter 4 for facts about correlation
Related Study Guide exercise – Exercise 4.33

Exercise 7.25
Text location – Chapter 6 for conditional distributions
Related Study Guide exercise – Exercise 6.5

Exercise 7.27
Text location – Chapter 3 for normal calculations
Related Study Guide exercises – Exercises 3.9, 3.14

Exercise 7.29
(a) Text location – Chapter 5 for the least-squares regression line
Related Study Guide exercise – Exercise 5.4

(b) Text location – Chapter 5 for the least-squares regression line
Related Study Guide exercise – Exercise 5.4

(c) Text location – Chapter 5 for regression lines
Related Study Guide exercise – Exercise 5.4

Exercise 7.31
(a) Text location – Chapter 4 for facts about correlation
Related Study Guide exercise – Exercise 4.32

(b) Text location – Chapter 4 for facts about correlation.
Related Study Guide exercise – Exercise 4.32

Exercise 7.33
Text location – Chapter 4 for facts about correlation
Related Study Guide exercise – Exercise 4.32

Exercise 7.35
Text location – Chapter 5 for association does not imply causation
Related Study Guide exercise – Exercise 5.47

CHAPTER 8

PRODUCING DATA: SAMPLING

OVERVIEW

Data can be produced in a variety of ways. **Observational studies** are investigations in which is simply observed the state of some population, usually with data collected by sampling. Even with proper sampling, data from observational studies are generally not appropriate for investigating cause-and-effect relations between variables. The reason is that the explanatory variable can be **confounded** with lurking variables, so its effects on the response cannot be distinguished from those of the lurking variables. **Experiments** are investigations in which data are generated by active imposition of some treatment on the subjects of the experiment. Properly designed experiments are the best way to investigate cause-and-effect relations between variables.

Sampling, when done properly, can yield reliable information about a **population**. The population is the entire group of individuals or objects about which we want information. The information collected is contained in a **sample**, a part of the population observed. How the sample is chosen (the **sampling design**) has a large impact on the usefulness of the data. A useful sample will be representative of the **population** and will help answer our questions. "Good" methods of collecting a sample include the following:

> **Simple random samples,** sometime denoted **SRS**
> **Probability samples**
> **Stratified random samples**
> **Multistage samples**

All these sampling methods involve some aspect of randomness through the use of a formal chance mechanism. Random selection is just one precaution that a person can take to reduce **bias,** the systematic favoring of a certain outcome. Samples we select using our own judgment or because they are convenient are usually biased. Hence, we use computers, calculators, or a **table of random digits** to help us select a sample.

A simple random sample, or SRS, of size n is a collection of n individuals chosen from the population in a manner so that each possible set of n individuals has an equal chance of being selected. In practice, a simple method such as drawing names from a hat is one way of getting an SRS. The names in the hat are the individuals in the population. To choose the sample, we mix up the names and select the sample of n at random from the hat. In reality the population may be very large and a computer, calculator, or a table of random numbers can be used to as an alternative to drawing the names from the hat.

The method of selecting an SRS using a table of random digits can be summed up in two steps.

  1. Give every individual in the population its own numerical label. All labels need to have the same number of digits.

2. Starting anywhere in the table (usually a spot selected at random), read off labels until you have selected as many labels as needed for the sample.

Another common type of sample design is a stratified random sample. First the population is divided into **strata** and then an SRS is chosen from each stratum. The strata are formed using some known characteristic of each individual thought to be associated with the response to be measured. Examples of strata are gender or age. Individuals in a particular stratum should be more like one another than those in the other strata.

Poor sample designs include the **voluntary response sample**, where people place themselves in the sample, and the **convenience sample**. Both these methods rely on personal decision for the selection of the sample, generally a guarantee of bias in the selection of the sample.

Other kinds of bias can occur even in well designed studies:

> **Nonresponse bias**, which occurs when individuals who are selected do not participate or cannot be contacted
> Bias in the **wording of questions**, lea ding the answers in a certain directio n
> **Confounding** or confusing the effect of two or more variables
> **Undercoverage**, which occurs when some group in the population is given either no chance or a much smaller chance than other groups to be in the sample
> **Response bias**, which occurs when individuals participate but are not responding truthfully or accurately due to the way the question is worded, the presence of an observer, fear of a negative reaction from the interviewer, or any other such source

These types of bias can occur even in a randomly chosen sample, and we need to try to reduce their impact as much as possible.

GUIDED SOLUTIONS

Exercise 8.1

KEY CONCEPTS: Explanatory and response variables, experiments and observational studies

What are the researchers trying to demonstrate with this study? What groups are being compared, and did the experiment deliberately impose membership in the groups on the subjects to observe their responses? What are the explanatory and response variables?

Explanatory variable:

Response variable:

Observational study or experiment (circle one)? Why?

Exercise 8.4

KEY CONCEPTS: Population, samples

(a) What population is of interest to the political scientist? What population is the sample from? Are the two populations the same in this case?

(b) How many individuals responded to the question? Those individuals make up the sample.

Exercise 8.10

KEY CONCEPTS: Selecting an SRS with a table of random numbers

The table of random numbers can be used to select an SRS of numbers. To use it to select a random sample of six minority managers, the managers need to be assigned numerical labels. So that everyone does the problem the "same" way, we have labeled the managers according to alphabetical order.

01 Abdulhamid	08 Duncan	15 Huang	22 Puri
02 Agarwal	09 Fernandez	16 Kim	23 Richards
03 Baxter	10 Fleming	17 Liao	24 Rodriguez
04 Bonds	11 Gates	18 Mourning	25 Santiago
05 Brown	12 Goel	19 Naber	26 Shen
06 Castillo	13 Gomez	20 Peters	27 Vega
07 Cross	14 Hernandez	21 Pliego	28 Wang

If you go to line 139 in the table and start selecting two-digit numbers, you should get the same answer as the one given in the complete solution. If your entire sample has not been selected by the end of line 139, continue to the next line in the table.

The sample consists of the managers:

Exercise 8.13

KEY CONCEPTS: Stratified random sample

What are the two strata from which you are going to sample? A stratified random sample consists of combining an SRS from each stratum to form the full sample. How large an SRS will be taken from each of the two strata? How would you label the units in each of the two strata from which you will sample? Fill in the following table to describe your sampling plan.

Strata

	Midsize accounts	Small accounts
Number of units in stratum		
Sample size		
Labeling method		

In practice, you would probably use the table of random numbers to first select the SRS from the midsize accounts and then select the SRS from the small accounts. In this problem you are going to select not the full samples but only the first five units from each stratum. Start in Table B at line 115 and first select 5 midsize accounts. Then continue in the table to select 5 small accounts. Write the numerical labels below.

First five midsize accounts:

First five small accounts:

Exercise 8.29

KEY CONCEPTS: Observational studies and experiments, lurking variables

(a) Records of over 850,000 operations were examined and both the anesthetic used and the surgical outcome were noted. Was the choice of anesthetic used made by those running the study? Why would we call this an observational study?

(b) Are there any obvious variables that might be associated with a doctor's choice of anesthetic? Suggest how these lurking variables could explain the relationship between the anesthetic and the death rate.

Exercise 8.31

KEY CONCEPTS: Populations and sources of bias

(a) Try to identify the population as exactly as possible from the information given. What is the sample size?

(b) If the order was not rotated, is it possible for bias to be introduced? Explain.

Exercise 8.44

KEY CONCEPTS: Systematic sampling

(a) This exercise is like the example except that there are now 200 addresses instead of 100, and the sample size is now 5 instead of 4. With these two changes, you need to think about how many different systematic samples there are. Two different systematic samples follow.

systematic sample 1 = 01, 41, 81, 121, 161
systematic sample 2 = 02, 42, 82, 122, 162

How many systematic samples are there altogether? Choosing one of these systematic samples at random is equivalent to choosing the first address in the sample. The remaining four addresses follow automatically by adding 40. Carry this calculation out using line 120 in the table.

(b) Why are all addresses equally likely to be selected? First, how many systematic samples contain each address? The chance of selecting an address is the same as the chance of selecting the systematic sample that contains it. With this in mind, what is the chance of any address being chosen? The definition of an SRS requires that all samples of 5 addresses are equally likely to be selected. In a systematic sample, are all samples of 5 addresses even possible?

Exercise 8.45

KEY CONCEPTS: Sampling frame, undercoverage

(a) Which households wouldn't be in the sampling frame? Make some educated guesses as to how these households might differ from those in the sampling frame (other than the fact that they don't have a phone number in the directory).

(b) Random digit dialing makes the sampling frame larger. Which households are added to it?

Exercise 8.46

KEY CONCEPTS: Wording of questions

A question can be worded in a way to make it seem as though any reasonable person should answer yes (or no). Which questions are slanted toward a desired response? Are all questions clear?

(a)

(b)

(c)

Exercise 8.49

KEY CONCEPTS: Populations, samples, sample size, bias

(a) What is the population of interest and the sample? Is the sample from the population of interest or are there potential sources of bias?

(b) Is this a scientific sample using probability in the selection of the sample? Is the sample size large?

COMPLETE SOLUTIONS

Exercise 8.1

The explanatory variable is whether or not the person made use of handheld cellular phones (we assume on a regular basis), and the response is whether or not the person contracted brain cancer. No attempt was made to decide which individuals were going to make use of a cellular phone (the treatment), so this is an observational study, not an experiment.

Exercise 8.4

(a) The population of interest to the political scientist is all college students. Unfortunately, for convenience the sample is selected from only the undergraduates attending her college. Suppose her college is a small liberal eastern college or a conservative evangelical college. In these cases, the opinions of students attending her college on a political issue such as social security may differ from opinions of the population of all college students, biasing the results.

(b) The sample is the 104 questionnaires returned. More than 50% of the 250 students to whom the questionnaire was mailed the did not respond, which can introduce nonresponse bias.

Exercise 8.10

To choose an SRS of six managers to be interviewed, first label the members of the population by associating a two-digit number with each.

01 Abdulhamid	08 Duncan	15 Huang	22 Puri
02 Agarwal	09 Fernandez	16 Kim	23 Richards
03 Baxter	10 Fleming	17 Liao	24 Rodriguez
04 Bonds	11 Gates	18 Mourning	25 Santiago
05 Brown	12 Goel	19 Naber	26 Shen
06 Castillo	13 Gomez	20 Peters	27 Vega
07 Cross	14 Hernandez	21 Pliego	28 Wang

Now go Table B inside the back cover of your textbook and read two-digit groups until 6 managers are chosen. Starting at line 139,

55|58|8 9|94|04| 70|70|8 4|10|98 43|56|3 5|69|34| 48|39|4 5|17|19| 12|97|5 1|32|58| 13|04|8

The selected sample is 04 Bonds, 10 Fleming, 17 Liao, 19 Naber, 12 Goel, and 13 Gomez.

Exercise 8.13

There are 500 midsize accounts. We are going to sample 5% of these, which is 25. You should label the accounts 001, 002, ⋯, 500 and select an SRS of 25 of the midsize accounts. There are 4400 small accounts. We are going to sample 1% of these, which is 44. You should label the accounts 0001, 0002, ⋯, 4400 and select an SRS of 44 of the small accounts.

Starting at line 115, we first select 5 midsize accounts, that is an SRS of size 5, using the labels 001 through 500. Continuing in the table we select 5 small accounts, that is, an SRS of size 5, using the labels 0001 through 4400. Note that for the midsize accounts we read from Table B using three-digit numbers, and for the small accounts we read from the table using four-digit numbers.

610|41 7|768|4 94|322| 247|09 7|3698| 1452|6 318|93 32|59 1|4459| 2605|6 314|24 80|371 6|

The first 5 midsize accounts are those with labels 417, 494, 322, 247, and 097. Continuing in the table, using four digits instead of three, the first 5 small accounts are those with labels 3698, 1452, 2605, 2480, and 3716.

Exercise 8.29

(a) The study looks at records that include both the outcome of the surgery and the anesthetic used. Although no mention is made of how the anesthetics were assigned to the patients, there are many possible variables that could have been used that may also have been related to the likelihood of a patient dying.

(b) The doctor's choice of anesthetic could be associated with the difficulty of the surgery, condition of the patient, or length of the surgery. If only long, difficult surgeries done on patients who are the most seriously ill used anesthetic C, then it would be no surprise that the death rate was higher.

Exercise 8.31

(a) The population of interest is probably adults aged 18 or older in the United States or possibly just registered voters. A possible source of bias is that only residents with phones could be contacted, and if the phone numbers were selected from phone books, then residents with unlisted numbers could not be in the sample. The sample size is the 1002 adults who agreed to be interviewed.

(b) When presented with four choices, if an individual has no particular opinion there may be a tendency to choose the first or last choice, for example. This kind of response would bias the results in favor of these choices if the responses were not rotated.

Exercise 8.44

(a) We want to select 5 addresses out of 200, so we think of the 200 addresses as 40 lists, each containing 5 addresses. We choose one address from the first 40 and then every 40th address after that. The first step is to go to Table B, line 120, and choose the first two-digit random number you encounter that is one of the numbers 01, $\cdots$, 40.

<u>35</u>476

The selected number is 35, so the sample includes addresses numbered 35, 75, 115, 155, and 195.

(b) Each individual is in exactly one systematic sample, and the systematic samples are equally likely to be chosen. In our previous example, there were 40 systematic samples, each containing 5 addresses. The chance of selecting any address is the chance of picking the systematic sample that contains it, which is 1 in 40.

A simple random sample of size n would allow every set of n individuals an equal chance of being selected. Thus, in this exercise, when using an SRS the sample consisting of the addresses numbered 1, 2, 3, 4, and 5 would have the same probability of being selected as any other set of 5 addresses. For a systematically selected sample, all samples of size n do not have the same probability of being selected. The sample consisting of the addresses numbered 1, 2, 3, 4, and 5 would have zero chance of being selected since the numbers of the addresses do not all differ by 40. The sample we selected in (a), 35, 75, 115, 155, and 195 had a 1 in 40 chance of being selected, so all samples of 5 addresses are not equally likely.

Exercise 8.45

(a) Households omitted from the frame are those that do not have a telephone number listed in the telephone directory. The types of people who might be underrepresented are poorer people (including the homeless) who cannot afford a phone and the group of people who have unlisted numbers. It is harder to characterize this second group. As a group they would tend to have more money as you need to pay to have your phone number unlisted or the group might include more single women who do not want their phone numbers available and possibly people whose jobs put them in contact with large groups of people who might harass them if their phone numbers were easily accessible.

(b) People with unlisted numbers will be included in the sampling frame. The sampling frame would now include any household with a phone. One interesting point is that all households would not have the same probability of being in the sample as some households have multiple phone lines and would be more likely to get into the sample. So, strictly speaking, random-digit dialing would actually provide not an SRS of households with phones but an SRS of phone numbers!

Exercise 8.46

(a) The beginning of the question suggests that cell phone use is associated with brain cancer. This initial suggestion and the wording "the danger of using cell phones" would lead most reasonable people to be in favor of including a warning label. The question is slanted in favor of this response.

(b) The question is clear but is slanted in favor of national health insurance. The reason for agreeing with a question should not be contained within the question.

(c) The question is slanted because it contains reasons for not supporting government subsidies for day care programs. As a question, the wording is a little technical and unclear, using phrases such as "negative externalities in parent labor force participation" and "increased group size with morbidity of children." A simpler, unslanted version such as "Do you support government subsidies for day care programs?" would be better.

Exercise 8.49

(a) The population is all people who live in Ontario. Because everyone uses health care, it is not restricted to adults, and so on The sample is the 61,239 residents of Ontario who were interviewed.

(b) Yes. This is a very large sample and it is a probability sample, so we expect that the sample proportions are quite close to the population proportions.

CHAPTER 9

PRODUCING DATA: EXPERIMENTS

OVERVIEW

Experiments are studies in which one or more **treatments** are imposed on experimental **units** or **subjects**. A treatment is a combination of **levels** of the explanatory variables, called **factors**. The design of an experiment is a specification of the treatments to be used and the manner in which units or subjects are assigned to these treatments. The basic features of well-designed experiments are **control**, **randomization**, and **replication**.

Control is used to avoid confounding (mixing up) the effects of treatments with other influences such as lurking variables. One such lurking variable is the **placebo effect**, which is the response of a subject to the fact of receiving any treatment. The simplest form of control is **randomized comparative experimentation**, which involves comparisons between two or more treatments. One of these treatments may be a **placebo** (fake treatment), and subjects receiving the placebo are referred to as a **control group**.

Randomization can be carried out using the ideas in Chapter 8 of your text. Randomization is carried out before applying the treatments and helps control bias by creating treatment groups that are similar. **Replication**, the use of many units in an experiment, is important because it reduces the chance variation between treatment groups arising from randomization. Using more units helps increase the ability of your experiment to establish differences between treatments.

Further control in an experiment can be achieved by forming experimental units into **blocks** that are similar in some way thought to affect the response, similar to strata in a stratified sample design. In a **block design**, units are first formed into blocks, and then randomization is carried out separately in each block. **Matched pairs** are a simple form of blocking used to compare two treatments. In a matched pairs experiment, either the same unit (the block) receives both treatments in a random order or very similar units are matched in pairs (the blocks). In the latter case, one member of the pair receives one of the treatments and the other member the remaining treatment. Members of a matched pair are assigned to treatments using randomization.

Some additional problems that can occur that are unique to experimental designs are **lack of blinding** and **lack of realism**. These problems should be addressed when designing the experiment.

GUIDED SOLUTIONS

Exercise 9.3

KEY CONCEPTS: Identifying experimental units, factors, treatments, and response variables

Read the description of the study carefully.   To identify the "individuals," ask yourself, what the treatments are going to be applied to. What is being measured on these individuals? This is the response.

Draw a diagram like that in Figure 9.1 of your text here. What are the factors and how many levels do they have? How many combinations of the levels of the factors are there? This number is the number of treatments.

Exercise 9.6

KEY CONCEPTS: Completely randomized design, randomization

How many treatment groups are there and how many rats are assigned to each group? This information must be included in your outline. Now, using Figure 9.3 of your text as a model, draw an appropriately labeled diagram that outlines a completely randomized design for the study.

Label the rats 01 to 18. At line 142 of Table B, select an SRS of 6 rats to receive black tea extract. Continue in Table B, selecting 6 more to receive green tea extract. The remaining 6 receive the placebo.

Six rats assigned to receive black tea extract:

Six rats assigned to receive green tea extract:

Six rats assigned to receive placebo:

Exercise 9.8

KEY CONCEPTS: Randomized comparative experiments, observational studies

Is the new design suggested by the executive an experiment? What are the disadvantages of the approach?

Exercise 9.9

KEY CONCEPTS: Randomized comparative experiments, observational studies

Is either of the designs observational studies? If so, what are some possible lurking variables that would produce less trustworthy data.

Exercise 9.12

KEY CONCEPTS: Double-blind experiments, bias

What does it mean for the ratings to be blind? Was this done in this case? If not, how can this bias the results and in which direction?

Exercise 9.13

KEY CONCEPTS: Matched pairs design, randomization

First identify the treatments and the response variable. Next, decide what makes up the matched pairs in this experiment. How will you use a coin flip to assign members of a pair to the treatments? What will you measure and how will you decide whether the right-hand tends to be stronger in right-handed people?

Exercise 9.32

KEY CONCEPTS: Design of an experiment, randomization

To begin, identify the subjects, the factors, the treatments, and the response variable. Now outline your design. Be sure to specify the following.

• How many treatments are there? Hence, how many groups of subjects must you form?

• How will you assign subjects to treatment groups?

• What are the treatments – what will each subject be required to do?

• What response will you measure and how will you decide whether the treatments differ in their effect?

You can outline your design in words or in a drawing.

Following is the list of names. Assign a numerical label to each one. Be sure to use the same number of digits for each label. So that everyone does the problem in the "same" way, we used the convention of starting with the label 00 for Acosta and continuing to label down the columns (01 Asihiro, etc). Of course, one could start with another number (such as 01) and label across rows if one wished.

Acosta	Farouk	Liang	Solomon
Asihiro	Fleming	Maldonado	Trujillo
Bennett	George	Marsden	Tullock
Bikalis	Han	Montoya	Valasco
Chen	Howard	O'Brian	Vaughn
Clemente	Hruska	Ogle	Wei
Duncan	Imrani	Padilla	Wilder
Durr	James	Plochman	Willis
Edwards	Kaplan	Rosen	Zhang

Now start at line 130 in Table B. Read across the row in groups of digits equal to the number of digits you used for your labels (if you used two digits for labels, read line 130 in pairs of digits). Keep reading until you have selected all the names for the first treatment. You may need to continue on to line 131, line 132, and subsequent lines. After you have selected the names for Treatment 1, continue in Table B to assign the nine people to receive Treatment 2 and then nine to receive Treatment 3. The remaining names are assigned to Treatment 4.

Exercise 9.36

KEY CONCEPTS: Matched pairs design, randomization

First identify the subjects, the treatments, and the response variable. Next decide what makes up the matched pairs in this experiment. How will you use the table of random numbers to assign members of a pair to the treatments? Why is it important to have each player's trials on different days?

The first 20 digits of Table B at line 170 are reproduced here. Use them to decide which players will get oxygen on their first trial.

38075 73239 52555 46342

COMPLETE SOLUTIONS

Exercise 9.3

The individuals are the participating schools as the treatments are applied to the school. The observed response includes the measure of physical activity and the lunchtime consumption of fat at each school.

There are two factors in the experiment — physical activity intervention and nutritional intervention. Two levels of physical activity intervention correspond to its use or nonuse in the school, and two levels of nutritional intervention correspond to its use or nonuse in the school, giving a total of four treatments. The diagram below shows the different treatment combinations in the design.

Factor B
Nutritional Intervention

		Yes	No
	Yes	1	2
Factor A Physical Activity Intervention			
	No	3	4

Exercise 9.6

There are three treatments, black tea extract, green tea extract, and a placebo. There are 18 rats for the study, so 6 rats are assigned at random to each of the treatment groups.

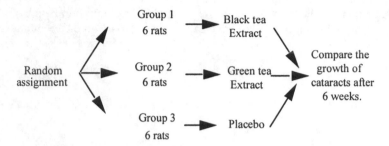

Table B starting at row 142 is reproduced here. Read across the row in groups of two digits since you used two digits for your labels, skipping repeats. The first six underlined numbers correspond to the rats receiving black tea extract and the next six to those receiving green tea extract. The remaining rats receive the placebo.

```
72829  50232  97892  63408  77919  44575  24870  04178
88565  42628  17797  49376  61762  16953  88604  12724
62964  88145  83083  69453  46109  59505  69680  00900
19687  12633  57857  95806  09931  02150  43163  58636
37609  59057  66967  83401  60705  02384  90597  93600
54973  86278  88737  74351  47500  84552  19909  67181
00694  05977  19664  65441  20903  62371  22725  53340
71546  05233  53946  68743  72460  27601  45403  88692
07511  88915  41267
```

Rats assigned to receive black tea extract	02, 08, 17, 10, 05 and 09
Rats assigned to receive green tea extract	06, 16, 01, 07, 18 and 15
Rats assigned to receive placebo	03, 04, 11, 12, 13, and 14

Exercise 9.8

The new design is an observational study, not an experiment. The electric company wants the only systematic differences in groups to be the treatments. Electric use varies from year to year depending on the weather. If charts or indicators are introduced in the second year, and the electric consumption in the first year is compared with the second year, you won't know if the observed differences are due to the introduction of the chart or to lurking variables. For example, if the comparison is being made in the summer months, it is possible that the second year had a cooler summer, which reduced the need for air conditioning and reduced electric consumption, rather than the introduction of charts or indicators. A control group ensures that influences other than the introduction of the indicators or charts operate equally on all groups.

Exercise 9.9

The first design is an observational study. The exercise treatment is not imposed on the groups, subjects have chosen whether or not to exercise. Other factors (confounding factors) may be related to the decision to exercise and the likelihood of a heart attack, such as personality type, eating habits, weight, smoking and so forth. We cannot conclude that exercise reduces the risk of a heart attack as the reduction in risk could be due to one of the confounding factors. Although matching of subjects is an attempt to control for these variables, it is not guaranteed to eliminate the effect of all confounding variables and is not a substitute for a well designed experiment. The second design is an experiment as each subject is randomly assigned to an exercise program or to continue with their "usual habits." If the subjects' "usual habits" include some sort of exercise, then this is fine: the two levels of the treatment are "to exercise" or "make your own decisions about exercise".

Exercise 9.12

The ratings were not blind because the experimenter who rated the subjects' level of anxiety presumably knew whether the subjects were in the meditation group or not. Since the experimenter was hoping to show that those who meditated had lower levels of anxiety, he might unintentionally rate those in the meditation group as having lower levels of anxiety, if there is any subjectivity in the ratings. It would be better if a third party who did not know which group the subjects belonged to rated the subjects' anxiety levels.

Exercise 9.13

We have 10 subjects. There are two treatments in the study. Treatment 1 is squeezing with the right hand and treatment 2 is squeezing with the left hand. The response is the force exerted as indicated by the reading on the scale.

To do the experiment, we use a matched pairs design. The matched pairs are the two hands of the subjects. We randomly decide which hand to use first, perhaps by flipping a coin. We measure the response for each hand and then compare the forces for the left and right hands over all subjects to see if there is a systematic difference between the two hands.

Exercise 9.32

In this case, the subjects are the 36 headache sufferers who have agreed to participate in the study. The two factors are antidepressant (placebo or antidepressant given) and stress management training (given or not given), and they form the four treatments:

Treatment 1: Antidepressant and no stress management training

Treatment 2: Placebo (no antidepressant) and no stress management training

Treatment 3: Placebo (no antidepressant) and stress management training

Treatment 4: Antidepressant and stress management training

The response variables are the number of headaches over the study period and some measure of the severity of these headaches. The problem does not specify how the severity is to be measured.

Subjects should be randomly assigned to treatments, with nine assigned to Treatment 1, nine assigned to Treatment 2, nine assigned to Treatment 3, and the remainder assigned to Treatment 4. Each subject follows his or her treatment regimen over the course of the study. The average number of headaches and the severity of the headaches for each treatment should be calculated and the results for the four groups compared. A drawing summarizing the experimental design is given on the next page.

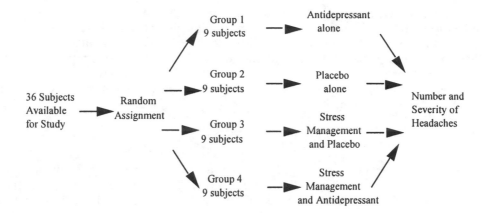

Although you can use the applet or Table B, we illustrate the use of Table B to carry out the random assignment of the subjects to the treatments. First label the 36 names using two-digit labels. We use the convention of starting with the label 00 and labeling down the columns. Of course, one could start with another number (such as 01) and label across rows, if one wished.

00 Acosta	09 Farouk	18 Liang	27 Solomon
01 Asihiro	10 Fleming	19 Maldonado	28 Trujillo
02 Bennett	11 George	20 Marsden	29 Tullock
03 Bikalis	12 Han	21 Montoya	30 Valasco
04 Chen	13 Howard	22 O'Brian	31 Vaughn
05 Clemente	14 Hruska	23 Ogle	32 Wei
06 Duncan	15 Imrani	24 Padilla	33 Wilder
07 Durr	16 James	25 Plochman	34 Willis
08 Edwards	17 Kaplan	26 Rosen	35 Zhang

Line 130 from Table B follows. We should read line 130 in pairs of digits from left to right. Vertical bars are placed between consecutive pairs to indicate how to read the table. The pairs that correspond to labels in the list and that have not been previously selected are underlined.

69|<u>05</u>|<u>16</u>|48|<u>17</u>|87|17|40|95|17|84|53|40|64|89|87|<u>20</u>|<u>19</u>|72|45

We find only 5 of our labels so we need to continue reading on line 131.

05|<u>00</u>|71|66|<u>32</u>|81|19|41|48|73|<u>04</u>|19|78|55|76|45|19|59|65|65

We now have 8 labels, and continuing to line 132 gives us the 9th label for the group assigned to Treatment 1.

68|73|<u>25</u>|52|59|84|29|20|87|96|43|16|59|37|39|31|68|59|71|50

The Treatment 1 group consists of Clemente, James, Kaplan, Marsden, Maldonado, Acosta, Wei, Chen, and Plochman. Continuing in line 132 (and ignoring pairs corresponding to previously selected subjects), we assign the next nine subjects to Treatment 2.

|52|59|84|<u>29</u>|20|87|96|43|16|59|37|39|<u>31</u>|68|59|71|50

45|74|04|<u>18</u>|<u>07</u>|65|56|<u>13</u>|<u>33</u>|<u>02</u>|07|05|19|36|<u>23</u>|18|13|20|95|47

<u>27</u>|

giving the Treatment 2 group as Tullock, Vaughn, Liang, Durr, Howard, Wilder, Bennett, Ogle, and Solomon. Continuing in line 134,

$$|81|67|84|16|18|32|92|13|37|\ \underline{35}|\underline{21}|33|77|41|\underline{04}|\underline{31}|\underline{26}|85|\underline{08}$$

$$66|92|55|56|58|39|\underline{10}|07|84|58|\underline{11}|\underline{20}|61|98|76|87|\underline{15}|13|\underline{12}|60$$

$$\underline{08}|\underline{42}|\underline{14}|$$

The Treatment 3 group is Zhang, Montoya, Rosen, Edwards, Fleming, George, Imrani, Han, and Hruska. The nine remaining subjects, Asihiro, Bikalis, Duncan, Farouk, O'Brian, Padilla, Trujillo, Valasco, and Willis, are assigned to Treatment 4. Using Table B can become tedious with a large number of subjects, so it is best to leave such calculations to a computer.

Exercise 9.36

We have 20 players available and there are two treatments. Treatment 1 is inhaling oxygen during the rest periods, and Treatment 2 is not inhaling oxygen between the rest periods. The response is the time on the final run.

To do the experiment, we use a matched pairs design. Each player is a block and each player will run 100 yards four times under both treatments, three in quick succession followed by a fourth time after a 3 minute rest. We should randomly decide which treatment to use first. For the players to have a chance to recover, we should allow sufficient time between the two trials (say, a day or two).

To assign which treatment comes first for each player, suppose the players are numbered from 1 to 20. Go to Table B, line 170, to decide which players get oxygen on their first trial. Use the following randomization. If the random digit is 0, 1, 2, 3, or 4, then the subject gets oxygen on the first trial. Otherwise, the subject gets oxygen on the second trial. The first 20 digits in line 170 follow, with the digits 0, 1, 2, 3, and 4 underlined.

$$\underline{38}0 75\ 7\underline{3}2\underline{3}9\ 5\underline{2}555\ \underline{4}6\underline{342}$$

Since positions 1, 3, 7, 8, 9, 12, 16, 18, 19, and 20 correspond to digits 0, 1, 2, 3, or 4, players numbered 1, 3, 7, 8, 9, 12, 16, 18, 19, and 20 receive oxygen during the rest periods on their first trial and no oxygen during their second trial. The remaining players get oxygen during the rest periods on their second trial and none during their first trial. In this example, 10 players get oxygen during the rest periods on their first trial and 10 get oxygen during the rest period on their second trial. When using a table of random numbers or software to generate random numbers, there is no guarantee that the groups will balance out exactly every time the randomization is done.

As an alternative, you could label the players 01 through 20. Begin on line 170 and select 10 players at random to be in the oxygen group. The remaining 10 would not get oxygen during the rest period. This would guarantee balance.

CHAPTER 10

INTRODUCING PROBABILITY

OVERVIEW

Probability is the mathematical study of random processes over a long series of repetitions. A process or phenomenon is called **random** if its outcome is uncertain. Although individual outcomes are uncertain, when the process is repeated a large number of times the underlying distribution for the possible outcomes begins to emerge. For any outcome, its **probability** is the proportion of times, or the relative frequency, with which the outcome would occur in a long series of repetitions of the process. It is important that these repetitions or trials be **independent** for this property to hold. By independent we mean that the outcome of one trial must not influence the outcome of any other.

You can study random behavior by carrying out physical experiments such as coin tossing or die rolling, or you can simulate a random phenomenon on the computer. Using the computer is particularly helpful in to considering a large number of trials. Randomness and independence are the keys to using the rules of probability.

The description of a random phenomenon begins with the **sample space, S,** which is the list of all possible outcomes. A set of outcomes is called an **event.** Once we have determined the sample space, a **probability model** tells us how to assign probabilities to the various events that can occur. There are four basic rules that probabilities must satisfy.

- Any probability is a number between 0 and 1. $P(A)$ means "the probability of A." If the probability is 0, the event will never occur. If the probability is 1, the event will always occur.

- All possible outcomes together must have probability 1.

- The probability that an event does not occur is 1 minus the probability that the event occurs. Using notation,

$$P(A) = 1 - P(\text{not } A)$$

- If two events have no outcomes in common, the probability that one or the other occurs is the sum of their individual probabilities. These events are **disjoint.** This is the addition rule for disjoint events, namely

$$P(A \text{ or } B) = P(A) + P(B)$$

In a sample space with a finite number of outcomes, probabilities are assigned to the individual outcomes and the probability of any event is the sum of the probabilities of the outcomes it contains. All outcomes must have a probability between 0 and 1; the sum of all probabilities must add up to 1.

Even a sample space with an infinite number of outcomes can have probabilities assigned to it. To assign probablilites, we use a density curve. (Refer to Chapter 3 of your textbook to refresh your memory.) The area under a density curve must be equal to 1. Probabilities are assigned to events as areas under the density curve. The normal distribution is the most useful of the density curves. **Normal distributions are probability models**.

A **random variable** is a variable whose value is a numerical outcome of a random phenomenon. The **probability distribution** of a random variable tells us about the possible values of the random variable and how to assign probabilities to these values. A random variable can be **discrete** or **continuous**. A **discrete random variable** has finitely many possible values. Its distribution gives the probability of each value. A **continuous random variable** takes all values in some interval of numbers. A **density curve** describes the distribution of a continuous random variable.

GUIDED SOLUTIONS

Exercise 10.1

KEY CONCEPTS: Probability as the proportion of times an event occurs in many repeated trials

The probability of any outcome of a random phenomenon is the proportion of times the outcome would occur in a very long series of repetitions. Restate this in terms of getting four of a kind in Texas Hold'em. For example, a large series of repetitions would mean a very large number of hands of Texas Hold'em in which you hold a pair in your hand.

Explain why this does not mean that if you play 1000 hands in which you hold a pair in your hand, exactly 88 will be four of a kind.

Exercise 10.5

KEY CONCEPTS: Sample space

One of the main difficulties encountered when describing the sample space is finding some notation to express ideas formally. Following the text, our general format is $S = \{\quad\}$, where a description of the outcomes in the sample space is included within the braces.

(a) You want to express that any number between 0 and 24 is a possible outcome. You would write

$$S = \{\text{all numbers between 0 and 24}\}$$

(b) We know that the amount of money in coins (not bills) that the student is carrying is 0 or larger and, because the smallest coin is a penny, that it must be a multiple of $0.01. However, it is not clear what the largest possible amount should be. If you say that the highest possible amount is $10.00 in coins, why not $10.01? The point is that the best we probably can do is say that the amount of money is more than $0.00 and take the sample space to be (in dollars).

S = {set of all numbers to two decimal places greater than or equal to zero}
 = {$0.00, $0.01, $0.02, $0.03, $0.04, . . .}

(c) For simplicity, assume that only "whole" letter grades are possible and not grades like B+ or C–. Also assume the "usual" letter grade scale. Write S in the space provided.

$S =$

(d) What are the possible responses a student can make? Ignore the possibility of a response like "I don't know." Write S in the space provided.

$S =$

Exercise 10.29

KEY CONCEPTS: Estimating probabilities as the proportion of times an event occurs in many repeated trials

Record the number of heads for your 50 spins.

The probability of an event is the proportion of times the event occurs in many repeated trials of a random phenomenon. To estimate the probability of heads when you spin a nickel, determine the proportion of times heads occurred in your 50 trials.

Estimate =

Exercise 10.39

KEY CONCEPTS: Probabilities in a finite sample space, legitimate probability models

(a) In a legitimate probability model, the probabilities of all possible outcomes in the sample space must add to 1. The six colors listed and the outcome "The vehicle you choose has a color other than the six listed" account for all possible outcomes. Add up the six probabilities given in the table. What must the probability be that the vehicle you choose has a color other than the six listed?

Sum of the six probabilities listed in the table =

Probability that the vehicle you choose has a color other than the six listed =

(b) The events "The randomly chosen car is silver" and "the randomly chosen car is white" are disjoint. What do the rules of probability tell us about the probability that the randomly chosen car is either silver or white? Using this information, what do the rules of probability tell us about the probability that the event "The randomly chosen car is either silver or white" does not occur?

Exercise 10.45

KEY CONCEPTS: Sample spaces for simple random sampling, probabilities of events

(a) It's easy to make the list: $S = \{$(Abby, Deborah), (Abby, Mei-Ling), etc.$\}$. There is no need to include both (Abby, Deborah) and (Deborah, Abby) in your list, because both refer to the same two individuals. Write down list.

(b) How many outcomes are there in the sample space in part (a)? If they are equally likely, what is the probability of each?

(c) How many outcomes in S include Mei-Ling? When the outcomes are equally likely, the probability of the event is just

$$\frac{\text{number of outcomes in the event}}{\text{number of outcomes in } S} =$$

(d) How many outcomes in S include neither of the two men?

Exercise 10.48

KEY CONCEPTS: Random numbers, density curve

(a) Recall that a discrete random variable has finitely many possible values. A continuous random variable takes all values in some interval of numbers. Which is true in this example?

(b) The total area underneath any density curve is 1. Since the density curve has constant height over the range 0 to 2, what must the height be to make the area under this curve equal to 1? Note that since the area under the density curve is a rectangular region, the formula for the area of a rectangle will be useful.

Now draw a graph of the density curve.

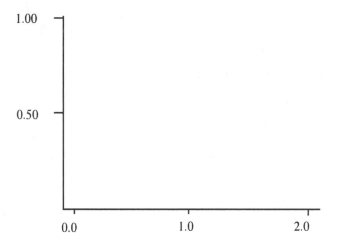

(c) $P(Y \leq 1)$ is the area under the density curve below 1. You may find it helpful to sketch this region on your graph in part (b) This is a rectangular region, so use the formula for the area of a rectangle to compute the area of this region.

Exercise 10.51

KEY CONCEPTS: Normal random variables

(a) Y stands for the score of a randomly chosen student. The event of interest is "The student has a score above 300." Because Y represents the student's score, in terms of Y the event is

$$\{Y > 300\}$$

To compute the probability of this event, recall that the mean is 300. What is the area under a normal curve corresponding to the region greater than the mean?

(b) Write the event in terms of Y. To compute the probability, first determine how many multiples of the standard deviation 35, 370 is above the mean? Now apply the 68–95–99.7 rule.

Exercise 10.55

KEY CONCEPTS: Simulating a random phenomenon

(a) You will need to use the applet or software to simulate the 100 trials. As the problem suggests, the key phrase to look for in your software is "Bernoulli trials." Your software may allow you to actually generate the words "Hit" and "Miss" or perhaps the letters H and M. However, your software is more likely to allow you to generate only the numbers 0 and 1. In this case, count a 0 as a miss and a 1 as a hit.

After you do so, the computer can be used to calculate the proportion of hits.

Proportion of hits =

For most students, the proportion of hits will be within 0.05 or 0.10 of the true probability of 0.5.

(b) You need to go through your sequence to determine the longest string of hits or misses.

Longest run of shots hit =   Longest run of shots missed =

COMPLETE SOLUTIONS

Exercise 10.1

To say that the probability of getting four of a kind is 88/1000 means that in a very large number of hands of Texas Hold'em in which you hold a pair in your hand, the proportion that will give you four of a kind is 88/1000.

The key phrase here is "a very large number of hands." "Very large" means much, much larger than 1000. Thus, a probability of 88/1000 means the proportion after millions and millions of hands in which you hold a pair, not after 1000.

Exercise 10.5

(a) S = {all numbers between 0 and 24}
(b) S = {set of all numbers to two decimal places greater than or equal to zero}
(c) S = {A, B, C, D, F}
(d) S = {Yes, No} or S = {Did take, Did not take}

Exercise 10.29

We spun a nickel 50 times and got 22 heads. From this result, we estimate the probability of heads to be

Estimate = 22/50 = 0.44

Your results will probably differ somewhat, but your estimate of the probability of heads will be the number of heads you got in 50 spins divided by 50.

Exercise 10.39

(a) The sum of the six probabilities listed in the table = 0.84.

The sum of all possible outcomes must be 1, so the probability that the vehicle you choose has a color other than the six listed when added to 0.84 must make the sum 1. Thus,

Probability the vehicle you choose has a color other than the six listed = 1 − 0.84 = 0.16

(b) Because the events are disjoint, we simply add the probabilities of silver and white to get

Probability that a randomly chosen vehicle is either silver or white = 0.18 + 0.17 = 0.35

Thus, by probability rule 4, namely that the probability that an event does not occur is 1 minus the probability that the event does occur,

Probability that a randomly chosen vehicle is neither silver nor white = 1 − 0.35 = 0.65.

Exercise 10.45

(a) S = {(Abby, Deborah), (Abby, Mei-Ling), (Abby, Sam), (Abby, Roberto), (Deborah, Mei-Ling), (Deborah, Sam), (Deborah, Roberto), (Mei-Ling, Sam), (Mei-Ling, Roberto), (Sam, Roberto)}.

(b) There are 10 possible outcomes. Since they are equally likely, each has probability 0.10.

(c) Mei-Ling is in four of the outcomes, so her chance of attending the conference in Paris is 4/10 = 0.4.

(d) The chosen group must contain two women, (Abby, Deborah), (Abby, Mei-Ling), or (Deborah, Mei-Ling). There are three possibilities, so the desired probability is 3/10 = 0.3.

Exercise 10.48

(a) Here Y takes on all values in the interval 0 to 2. Thus, Y is continuous.

(b) The density curve will be a rectangle with base covering the region 0 to 2. The base has length 2. The area of a rectangle is base × height, so the area of the density curve is 2 × height. This area must equal 1, so the height must be 0.5.

A graph of the density curve follows.

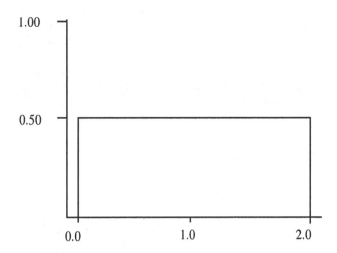

(c) Here is a graph of the desired region.

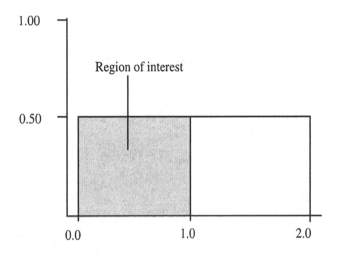

The region of interest is a rectangle with base of length 1.0 and height of 0.5. Thus the area of this rectangle is $1 \times 0.5 = 0.5$, so

$$P(Y \leq 1) = 0.5$$

Exercise 10.51

(a) In terms of Y, the event is $\{Y > 300\}$. Recall that half the area under the normal curve corresponds to the region greater than the mean and half to the region less than the mean. The mean in this case is 300, so

$$P(Y > 300) = \text{area under a normal curve greater than the mean} = 0.5$$

(b) In terms of Y, the event is $\{Y > 370\}$. Notice that 370 is 70 above 300. Because 70 is equal to twice the standard deviation of 35, 370 is two standard deviations above the mean. The 68–95–99.7 rule tells us that 95% of the area under a normal curve is within two standard deviations of the mean. Because of the symmetry of the normal curve, this statement implies that 2.5% of the area is more than two standard deviations below the mean and 2.5% of the area is more than two standard deviations above the mean. Thus,

$$P(Y > 370) = \text{probability of being more than two standard deviations above the mean} = 0.025$$

Exercise 10.55

(a) Here is our sequence of hits (H) and misses (M).

```
H   H   M   H   H   H   M   M   H   H   H   M   H   M   H
M   H   M   M   H   M   M   H   H   H   M   M   H   H   H
M   M   M   M   H   M   M   H   H   H   H   M   H   H   H
M   M   M   M   M   H   M   H   H   M   H   M   H   M   M
H   H   H   H   M   H   M   M   M   M   H   H   M   H   H
M   M   H   H   H   M   M   H   H   M   M   H   M   H   M
M   H   M   H   H   M   H   H   H   H
```

Proportion of hits = .54

(b) Go through your sequence to determine the longest string of hits or misses. In our example,

Longest run of shots hit = 4 (occurred more than once)

Longest run of shots missed = 5

CHAPTER 11

SAMPLING DISTRIBUTIONS

OVERVIEW

Statistical inference is the technique that allows us to use the information in a sample to draw conclusions about the population. Associated with any sample statistic is its **sampling distribution,** the distribution of values taken by the statistic in all possible samples of the same size from the same population. The sampling distribution can be described in the same way as the distributions you encountered in Chapters 1 and 2 of your textbook, and its three important features follow.

- Measure of center

- Measure of spread

- Description of the shape of the distribution

The statistic discussed first is the **sample mean**, $\bar{x}$, which is an estimate of the **population mean** μ. $\bar{x}$ has a number of convenient properties when taken from an SRS that allow one to make a variety of inferences about μ. As with all sampling distributions, we need to know the mean, standard deviation, and the shape of the distribution. In fact, we know from the **central limit theorem** that the shape of the distribution of the sample mean is very close to normal when simple random sampling is used and sample sizes are large. The **law of large numbers** also tells us more about the behavior of $\bar{x}$ as the sample size increases. The law of large numbers describes how the mean of many observations of a random process gets closer and closer to the mean of the population.

Here are the basic facts about the sample mean from an SRS of size n taken from a population where the mean is μ and the standard deviation is σ.

- $\bar{x}$ is an **unbiased estimate** of the population mean μ, so the mean of $\bar{x}$ is the population mean μ.

- The standard deviation is $\sigma/\sqrt{n}$, where σ is the population standard deviation. (There is less variation in averages than in individuals, and averages based on larger samples are less variable than those based on smaller samples.)

- The shape of the distribution of the sample mean depends on the shape of the population distribution. If the population was normal, $N(\mu, \sigma)$, then the sample mean has normal distribution $N(\mu, \sigma/\sqrt{n})$. Also, for a large sample the central limit theorem tells us that the sample mean has *approximately* a normal distribution $N(\mu, \sigma/\sqrt{n})$.

A statistic is called **unbiased** if the sampling distribution of the statistic is centered (has its mean) at the value of the population parameter. This means that the statistic tends to neither overestimate nor underestimate the parameter.

Statistical process control consists of methods for monitoring a process over time so that any changes in the process can be detected and corrected quickly. This is an economical method for maintaining product quality in a manufacturing process. We say that a process that continues over time is in **control** if it is operating under stable conditions. More precisely, the process is in control with respect to some variable measured on the process if the distribution of this variable remains constant over time.

Control charts are used to monitor the values of a variable measured on a process over time. One of the most common control charts is the $\bar{x}$ **control chart**. This chart is produced by periodically observing a sample of n values of the variable of interest and plotting the means $\bar{x}$ of these values versus the time order of the samples on a graph. A solid **centerline** at the target value of the process mean μ for the variable is drawn on the graph, as are dashed **control limits** at

$$\mu \pm 3 \frac{\sigma}{\sqrt{n}}$$

where σ is the process standard deviation of the variable. This chart helps us decide whether the process is in control with mean μ and standard deviation σ. The probability that the next point (value of $\bar{x}$) on such a chart lies outside the control limits is about 0.003 if the process is in control. Such a point would be evidence that the process is **out of control**, that is, that the distribution of the process has changed for some reason. When a process is deemed out of control, a cause for the change in the process should be sought.

GUIDED SOLUTIONS

Exercise 11.3

KEY CONCEPTS: Statistics and parameters

In deciding whether a number represents a parameter or a statistic, you need to think about whether it is a number that describes a population of interest or whether it is a number computed from the particular sample that was selected. What is the population and what is the sample in the problem? Based on the answer, indicate whether the number is a parameter or a statistic.

2.5003 cm:

2.5009 cm:

Exercise 11.8

KEY CONCEPTS: Standard deviation of the sampling distribution of $\bar{x}$

(a) The key formula is that if $\bar{x}$ is the mean of an SRS of size n drawn from a large population with mean μ and standard deviation σ, then the standard deviation of the sampling distribution of $\bar{x}$ is $\frac{\sigma}{\sqrt{n}}$. To apply this formula here, identify σ and n and then complete the following:

$$\text{Standard deviation of Juan's mean result} = \frac{\sigma}{\sqrt{n}} =$$

(b) What value would n have to be so that $\frac{\sigma}{\sqrt{n}}$ is 5?

Write out your explanation of the advantage of reporting the average of several measurements rather than the result of a single measurement. Remember, don't use technical language.

Exercise 11.13

KEY CONCEPTS: Central limit theorem

The four-step process follows.

   State. What is the practical question in the context of the real-world setting?

   Formulate. What specific statistical operations does this problem call for?

   Solve. Make the graphs and carry out any calculations needed for this problem.

   Conclude. Give your practical conclusion in the setting of the real-world problem.

To apply the steps to this problem, here are some suggestions.

State. What assumption about the average loss are you asked to investigate?

Formulate. You will need to assume that the 10,000 policies sold belong to homeowners who can be safely assumed to be an SRS from the population of all homeowners. What probability about the average loss do you need to calculate?

Solve. What does the central limit theorem say about the sampling distribution of the average loss, $\bar{x}$, for 10,000 policies? Fill in the blanks.

Sampling distribution of $\bar{x}$ is approximately $N($, $)$

Now use the normal probability calculations you learned in Chapter 3 to compute the probability that the mean $\bar{x}$ is no greater than \$275. If you have forgotten how to do these calculations, review in Chapter 3.

Conclude. What does the probability you calculated in the Solve step tell you about whether the company can safely base its rates on the assumption that the average loss will be no greater than \$275? In your opinion, is the probability sufficiently large?

Exercise 11.28

KEY CONCEPTS: Law of large numbers

Review the statement of the law of large numbers in the text. The mean payoff on a $1.00 bet is $0.947. What are the mean winnings for the gambler (μ in this problem)? The amount a gambler makes per bet on average after many bets on red would be $\bar{x}$. What does the law of large numbers say about $\bar{x}$?

Exercise 11.43

KEY CONCEPTS: Sampling distributions, central limit theorem

(a) Remember to read line 101 in groups of two digits at a time. What are the first five different two-digit numbers in line 101? Which of the corresponding circles are white? Thus,

$$\hat{p} =$$

(b) List the numbers selected and the corresponding values of $\hat{p}$, the proportion of white circles (those who approve of gambling), by filling in the blanks.

line 102:   number of white circles = $\hat{p} =$

line 103:   number of white circles = $\hat{p} =$

line 104:   number of white circles = $\hat{p} =$

line 105:   number of white circles = $\hat{p} =$

line 106:   number of white circles = $\hat{p} =$

line 107:   number of white circles = $\hat{p} =$

line 108:   number of white circles = $\hat{p} =$

line 109:   number of white circles = $\hat{p} =$

line 110:   number of white circles = $\hat{p} =$

(c) Use the following axes for your histogram.

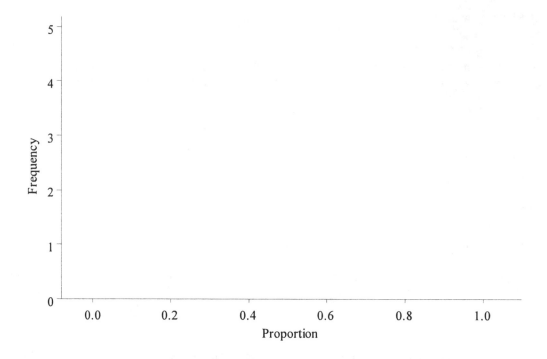

(d) The number of samples you obtained with $\hat{p} = 0.6$ is

Is the true value 0.6 roughly in the center of your histogram?

Exercise 11.45

KEY CONCEPTS: $\bar{x}$ control chart

Identify

$\mu =$

$\sigma =$

$n =$ sample size $=$

From this information, determine

Center line $= \mu =$

Control limits $= \mu \pm 3 \dfrac{\sigma}{\sqrt{n}} =$

Exercise 11.47

KEY CONCEPTS: Natural tolerances

Identify

$$\mu =$$
$$\sigma =$$

From this information, determine

Natural tolerance $= \mu \pm 3\sigma =$

COMPLETE SOLUTIONS

Exercise 11.3

We want information about the entire carload lot of ball bearings. The inspector chooses and tests 100 bearings from the lot to decide whether to accept or reject the entire lot. In this problem, the entire carload lot is the population and the 100 bearings the sample. Thus

2.5003 cm = a parameter since it describes the entire carload lot of bearings (the population)

2.5009 cm = a statistic since it describes the sample of 100 bearings

Exercise 11.8

(a) In this exercise $\sigma = 10$ and $n = 3$. Thus

$$\text{Standard deviation of Juan's mean result} = \frac{\sigma}{\sqrt{n}} = \frac{10}{\sqrt{3}} = 5.77$$

(b) We want $5 = \dfrac{\sigma}{\sqrt{n}} = \dfrac{10}{\sqrt{n}}$. Solving this equation for $\sqrt{n}$, we must have

$$\sqrt{n} = 10/5 = 2$$

Squaring both sides yields $n = 4$.

Averages of several measurements are more likely than a single measurement to be closer to the true value of the quantity being measured. The magnitude of chance deviations or errors are smaller for averages than for individual observation.

Exercise 11.13

State. You are asked to decide whether, if the company sells 10,000 policies, it can safely base its rates on the assumption that its average loss will be no greater than $275.

Formulate. We need to calculate the probability that the average annual loss, $\bar{x}$, for these 10,000 policies is no greater than $275.

Solve. We are told that the population of all homeowners has a mean annual loss of $250 and the standard deviation of the loss is $1000. The central limit theorem tells us that the sampling distribution of $\bar{x}$ is approximately normal with mean $250 (the same as the population mean) and standard deviation

$$\frac{\sigma}{\sqrt{n}} = \frac{\$1000}{\sqrt{10,000}} = \frac{\$1000}{100} = \$10$$

Thus,

Sampling distribution of $\bar{x}$ is approximately $N(\$250, \$10)$

The probability that $\bar{x}$ is no greater than $275 is

$$P(\bar{x} \le \$275) = P(\frac{\bar{x} - \$250}{\$10} \le \frac{\$275 - \$250}{\$10}) = P(z \le 2.5) = 0.9938$$

according to our normal tables.

Conclude. The probability that its average loss will be no greater than $275 is 0.9938. This probability is very close to 1. How close to 1 the probability needs to be so that the company can safely base its rates on the assumption that the average loss will be no greater than $275 is subjective. However, 0.9938 is probably large enough to be considered safe.

Exercise 11.28

The gambler pays $1.00 for an expected payout of $0.947. His mean winnings are therefore $0.947 – $1.00 = –$0.053 per bet. In other words, the expected losses of the gambler are $0.053 per bet. The law of large numbers tells us that if the gambler makes a large number of bets on red, keeps track of his net winnings, and computes the average $\bar{x}$ of these, this average will be close to –$0.053. In other words, he will find that he loses about 5.3 cents per bet on average.

Exercise 11.43

(a) The first five different two-digit numbers in line 101 are 19, 22, 39, 50, 34. The three circles labeled 39, 50, and 34 are white. Thus,

$$\hat{p} = 3/5 = 0.6.$$

(b) line 102: 73, 67, 64, 71, 50	number of white circles = 2	$\hat{p} = 2/5 = 0.4$
line 103: 45, 46, 77, 17, 09	number of white circles = 3	$\hat{p} = 3/5 = 0.6$
line 104: 52, 71 13, 88, 89	number of white circles = 2	$\hat{p} = 2/5 = 0.4$
line 105: 95, 59, 29, 40, 07	number of white circles = 0	$\hat{p} = 0/5 = 0.0$
line 106: 68, 41, 73, 50, 13	number of white circles = 3	$\hat{p} = 3/5 = 0.6$
line 107: 82, 73, 95, 78, 90	number of white circles = 1	$\hat{p} = 1/5 = 0.2$
line 108: 60, 94, 07, 20, 24	number of white circles = 4	$\hat{p} = 4/5 = 0.8$
line 109: 36, 00, 91, 93, 65	number of white circles = 4	$\hat{p} = 4/5 = 0.8$
line 110: 38, 44, 84, 87, 89	number of white circles = 3	$\hat{p} = 3/5 = 0.6$

(c) The histogram follows.

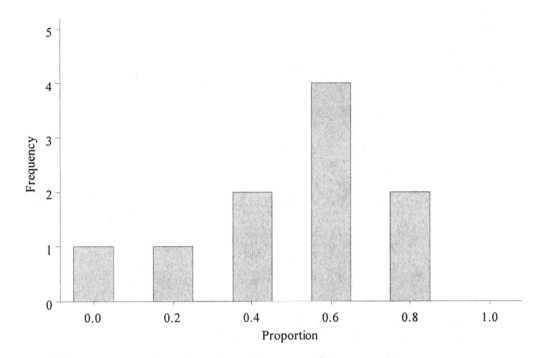

(d) The number of samples we obtained with $\hat{p} = 0.6$ is 4. The true value of 0.6 is roughly in the center of the histogram. It is certainly the most frequently occurring value.

Exercise 11.45

In this problem, $\mu = 4.22$
$\sigma = 0.127$
$n = \text{sample size} = 5$

From this information we find

Center line $= \mu = 4.22$

Control limits $= \mu \pm 3\,\dfrac{\sigma}{\sqrt{n}} = 4.22 \pm 3\,\dfrac{0.127}{\sqrt{5}} = 4.22 \pm 0.17$

Hence the lower control limit is 4.05 and the upper control limit is 4.39.

Exercise 11.47

We have $\mu = 4.22$
$\sigma = 0.127$

so

Natural tolerances $= \mu \pm 3\sigma = 4.22 \pm 3(0.127) = 4.22 \pm 0.38$.

CHAPTER 12

GENERAL RULES OF PROBABILITY

Overview

Chapter 6 of your text discusses two-way tables and conditional distributions. Here we learn about **conditional probabilities** and their use in calculating probabilities of complex events. The conditional probability of an event B given that an event A has occurred is denoted $P(B \mid A)$ and is defined by

$$P(B \mid A) = \frac{P(A \text{ and } B)}{P(A)}$$

where $P(A) > 0$. In practice, a conditional probability can often be determined directly from the information given in a problem. Events are **independent** if knowledge that one event has occurred does not alter the probability that the second event occurs. Specifically, two events A and B are independent if $P(B \mid A) = P(B)$. It follows that for independent events we must have the result that $P(A \text{ and } B) = P(A)P(B)$. In any particular problem we can use this result to check if two events are independent by determining if the probabilities multiply correctly. However, independence is usually assumed as part of the probability model.

The following general rules are valid for any assignment of probabilities and allow us to compute the probabilities of events in many random phenomena.

Addition rule: If events A, B, C are all disjoint in pairs, then

$$P(\text{at least one of these events occur}) = P(A) + P(B) + P(C)$$

Multiplication rule: If events A, B, C are independent, then

$$P(\text{all of the events occur}) = P(A)P(B)P(C)$$

General addition rule: For any two events A and B,

$$P(A \text{ or } B) = P(A) + P(B) - P(A \text{ and } B)$$

General multiplication rule: For any two events A and B,

$$P(A \text{ and } B) = P(A)P(B|A)$$

GUIDED SOLUTIONS

Exercise 12.3

KEY CONCEPTS: Independence, multiplication rule

What is the probability that a single textbook would have an author whose name was not among the 10 most common names?

What is the probability that all 9 textbooks would have an author whose name was not among the 10 most common names?

Exercise 12.4

KEY CONCEPTS: Independence, multiplication rule

The key concept that must be properly understood to answer the questions raised in this exercise is the notion of independence. Let A be the event an employed person has four years of college and B be the event that an employed person is a construction worker. Events A and B are independent if knowledge that A has occurred (person has four years of college) does not alter our assessment of the probability that B will occur (person is a construction worker).

What assumption has been made to use the multiplication rule to find the probability that A and B occur? Do you think the multiplication rule applies here?

Exercise 12.5

KEY CONCEPTS: Venn diagrams

Let C be the event that a college student likes country music and G be the event that a college student likes gospel music. First write the three probabilities provided in terms of C and G and then add these results to the Venn diagram that follows. When you are done, your Venn diagram should look similar to the Venn diagram in Figure 12.4 of your text.

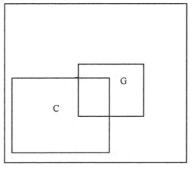

(a) You should be able to read this probability easily off the Venn diagram. This event corresponds to the portion of C that doesn't overlap with G.

(b) Students who like neither country nor gospel music are included in the region of the Venn diagram outside of the shapes corresponding to country and gospel.

Exercise 12.7

KEY CONCEPTS: Conditional probability

Let C be the event that a college student likes country music and G be the event that a college student likes gospel music. You are asked to find the conditional probability that a student likes gospel music given that he or she likes country music, namely, $P(G\,|\,C)$. Applying the general formula for conditional probabilities,

$$P(G\,|\,C) = \frac{P(G \text{ and } C)}{P(C)} =$$

Use the information from Exercise 12.5 to first evaluate the probabilities $P(G \text{ and } C)$ and $P(C)$. Then use these probabilities to compute the conditional probability above.

Exercise 12.30

KEY CONCEPTS: Multiplication rule for independent events

(a) The three years are independent. If U indicates a year for the price being up and D indicates a year for the price being down, you need to compute $P(UUU)$.

(b) This problem must be set up carefully and done in steps.

Step 1. Write the event of interest in terms of simpler outcomes. How would you write $P(UUU \text{ or } DDD)$ in terms of $P(UUU)$ and $P(DDD)$?

$P(\text{moves in the same direction in the next two years}) = P(UUU \text{ or } DDD) =$

Step 2. Evaluate $P(UUU)$ and $P(DDD)$ and substitute your answer in the expression from Step 1.

Exercise 12.38

KEY CONCEPTS: Conditional probabilities, geometric probabilities

You want to use geometric arguments to evaluate

$$P(Y < 1/2 | Y > X) = \frac{P(Y < 1/2 \text{ and } Y > X)}{P(Y > X)}$$

In the previous equation, we have just used the definition of conditional probability where A corresponds to "$Y < 1/2$" and B corresponds to "$Y > X$." The three figures that follow are of the square $0 \le x \le 1$ and $0 \le y \le 1$. In the figure on the left, the region corresponding to "$Y > X$" is shaded. Remembering that probabilities correspond to areas, first evaluate $P(Y > X)$.

$P(Y > X) =$

In the middle figure the region corresponding to "$Y < 1/2$" is shaded, and in the figure on the right the region "$Y < 1/2$ and $Y > X$" is shaded. The area of the shaded triangle in the figure on the right corresponds to the probability $P(Y < 1/2 \text{ and } Y > X)$. The value of this probability is

$P(Y < 1/2 \text{ and } Y > X) =$

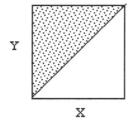 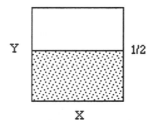 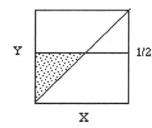

Putting this all together, you should be able to evaluate the following probability.

$$P(Y < 1/2 | Y > X) = \frac{P(Y < 1/2 \text{ and } Y > X)}{P(Y > X)} =$$

Exercise 12.41

KEY CONCEPTS: Two-way table of counts, conditional probabilities

Following is the table from the exercise.

	Bachelor's	Master's	Professional	Doctorate	Total
Female	645	227	32	18	922
Male	505	161	40	26	732
Total	1150	388	72	44	1654

(a) You can calculate this probability directly from the table. All degree recipients in the table are equally likely to be chosen (that is what it means to choose a degree recipient at random), so the fraction of the degree recipients in the table who are women is the desired probability. How many women degree recipients are there? Where do you find this number in the table? What is the total number of degree recipients represented in the table? Use these numbers to compute the desired fraction.

P(choose a woman) =

(b) This probability can also be calculated directly from the table. Since this is a conditional probability (this is a probability given that the degree recipient is a professional), we restrict ourselves only to professional degree recipients. The desired probability is then the fraction of these professional degree recipients who are women. Find the appropriate entries in the table to compute this fraction.

P(choose a woman | choose a professional degree recipient) =

(c) If the two events "choose a woman" and "choose a professional degree recipient" are independent, then what should be true about the probabilities computed in (a) and (b)? Are these events independent?

Exercise 12.49

KEY CONCEPTS: Multiplication rules and conditional probability, tree diagrams

We are given the following probabilities.

$$P(\text{pump regular gas}) = 0.40$$

$$P(\text{pump midgrade gas}) = 0.35$$

$$P(\text{pump premium gas}) = 0.25$$

$$P(\text{pay at least \$30} \mid \text{pump regular gas}) = 0.30$$

$$P(\text{pay at least \$30} \mid \text{pump midgrade gas}) = 0.50$$

$$P(\text{pay at least \$30} \mid \text{pump premium gas}) = 0.60$$

Draw a tree diagram as in Figure 12.5 to organize the information and to compute P(pay at least \$30). If you are having difficulty, this exercise is very similar to Example 12.9 in your text.

Exercise 12.51

KEY CONCEPTS: Conditional probability

We want to calculate the conditional probability P(pump premium gas | pay at least $30). Using the general formula for conditional probability we have

$$P\text{(pump premium gas | pay at least \$30)} = \frac{P\text{(pump premium gas and pay at least \$30)}}{P\text{(pay at least \$30)}}$$

In Exercise 12.49, you computed P(Pay at least $30). Use the tree diagram in Exercise 12.49 or the general multiplication rule to evaluate the numerator

P(pump premium gas and pay at least $30) =

You can now use these two probabilities to compute the desired conditional probability

P(pump premium gas | pay at least $30) =

Exercise 12.57

KEY CONCEPTS: Independence, multiplication rule

(a) The possible alleles inherited are B and B, B and O, and O and O. What blood types do these inherited alleles result in?

(b) Let S_O and S_B correspond to the events that allele O or B is inherited from Sarah, respectively, and D_O and D_B correspond to the events that allele O or B is inherited from David. S_O and S_B each has probability 0.5; D_O and D_B each has probability 0.5.

P(child has type O) = $P(S_O$ and $D_O)$ =

What rule allows you to multiply the probabilities?

How many blood types can their children have? What is the probability that the child has type B blood?

COMPLETE SOLUTIONS

Exercise 12.3

The probability that a single textbook would have an author whose name was not among the 10 most common names is $1 - 0.056 = 0.944$, assuming that a person's name is not related to whether or not they write a textbook.

The probability that all 9 textbooks would have an author whose name was not among the 10 most common names has probability

P(all 9 textbooks have authors whose name is not among the 10 most common) $= (0.944)^9 = 0.5953$

This probability assumes that the names of the authors of different textbooks are independent of each other. Thus it wouldn't be that surprising if none of the names of these authors were among the 10 most common.

Exercise 12.4

Suppose the events "college-educated" and "construction worker" are independent. This would imply that knowing whether someone was college-educated would not change the probability that the person was a construction worker. In terms of this problem, if the events were independent, 6% of employed people would be construction workers and 6% of the college-educated labor force would be construction workers (knowledge of whether or not an individual has four years of college doesn't alter (increase or decrease) their chance of being a construction worker). The independence is what allows us to just multiply these probabilities together. The use of the formula $(0.28)(0.06) = 0.017$ to get the answer requires that 6% of the college students are construction workers. However, we would guess that fewer than 6% of those with four years of college are construction workers so that multiplying the two probabilities together is not the correct way to get the answer. The answer obtained by multiplying the two probabilities together is too large.

Exercise 12.5

We are told that $P(C) = 0.4$, $P(G) = 0.3$, and $P(C \text{ and } G) = 0.10$. This information is reported in the Venn diagram. Since 40% of students like country and 10% like both country and gospel, we must have $0.4 - 0.1 = 0.3$ in the region corresponding to country and not gospel. Since 30% of students like gospel and 10% like both gospel and country, we must have $0.3 - 0.1 = 0.2$ in the region corresponding to gospel and not country. We have also entered 0.1 in the region corresponding to both country and gospel. Finally, the region outside the shapes corresponding to country and gospel must have 40% of the students since the four regions in the Venn diagram are disjoint and make up the entire sample space.

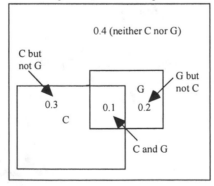

(a) From the Venn diagram we see that 30% of students like country but not gospel.

(b) From the Venn diagram we see that 40% of students like neither country nor gospel.

Exercise 12.7

From Exercise 12.5 we know that $P(G \text{ and } C) = 0.1$ and $P(C) = 0.4$. Plugging these values into the formula for conditional probability yields

$$P(G|C) = \frac{P(G \text{ and } C)}{P(C)} = \frac{0.1}{0.4} = 0.25$$

Exercise 12.30

(a) $P(UUU) = (0.65)^3 = 0.2746$

(b) Because the events UUU and DDD are disjoint

$P(\text{moves in the same direction in the next two years}) = P(UUU \text{ or } DDD) = P(UUU) + P(DDD).$

From part (a), $P(UUU) = 0.2746$. The probability of the price being down in any given year is $1 - 0.65 = 0.35$. Since the years are independent, the probability of the price being down in three consecutive years is $P(DDD) = (0.35)^3 = 0.0429$. Putting this together,

$P(\text{moves in the same direction in the next two years}) = 0.2746 + 0.0429 = 0.3175$

Exercise 12.38

The figure on the left in the guided solution has the region corresponding to "$Y > X$" shaded, and because probabilities correspond to areas we see that $P(Y > X) = 1/2$. In the figure on the right, the region "$Y < 1/2$ and $Y > X$" is shaded. The area of the shaded triangle in the figure on the right is 1/8, so $P(Y < 1/2 \text{ and } Y > X) = 1/8$.

Putting this all together gives

$$P(Y < 1/2 | Y > X) = \frac{P(Y < 1/2 \text{ and } Y > X)}{P(Y > X)} = \frac{1/8}{1/2} = 0.25$$

Exercise 12.41

(a) The number of women degree recipients is found as the total for the first row and is (in thousands) 922. The total number of degree recipients in the table is in the lower right corner and is (in thousands) 1654. The desired probability is thus

$P(\text{choose a woman}) = (\text{number of women degree recipients}) / (\text{total number of recipients in table})$

$$= 922/1654 = 0.5574$$

(b) The desired conditional probability is

P(choose a woman | choose a professional degree recipient)

= (number of professional degree recipients who are women) / (number of professional degree recipients)

= 32/72 = 0.4444

(c) If the two events "choose a woman" and "choose a professional degree recipient" are independent, then we should have

P(choose a woman) = P(choose a woman | choose a professional degree recipient)

These are the two probabilities that you computed in (a) and (b). Since they are not equal, these two events are not independent.

Exercise 12.49

The tree diagram below organizes the information given in the problem.

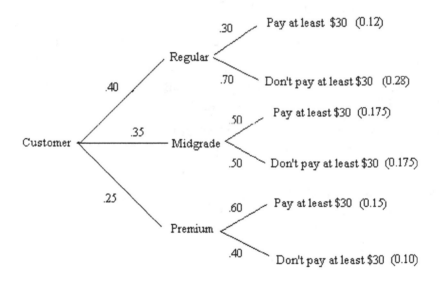

A customer buys regular, midgrade, or premium gas. The probabilities of purchasing these three types of gas mark the three leftmost branches in the tree. Look at the top branch corresponding to the purchase of regular gas. The two segments going out from the "regular" branch point have the conditional probabilities.

P(pay at least $30 | purchase regular) = 0.30

P(don't pay at least $30 | purchase regular) = 0.70

Now use the multiplication rule to find the probability that a purchaser of regular gas pays at least $30.

P(pay at least $30 and purchase regular) = P(purchase regular) P(pay at least $30 | purchase regular)

= (0.40)(0.30) = 0.12

This probability appears at the end of the topmost branch. The probabilities of all six complete branches are computed in this manner. There are three paths leading to "pay at least $30", and these paths are disjoint. Thus the probability of paying at least $30 is the sum of the probabilities associated with these three disjoint paths

P(pay at least $30) = 0.120 + 0.175 + 0.150 = 0.445.

Exercise 12.51

The conditional probability of interest is

$$P(\text{pump premium gas} \mid \text{pay at least \$30}) = \frac{P(\text{pump premium gas and pay at least \$30})}{P(\text{pay at least \$30})}$$

From the tree diagram, the event "pump premium gas and pay at least $30" is next to the last branch at the bottom of the tree and has

P(pump premium gas and pay at least $30)

$$= P(\text{pay at least \$30} \mid \text{pump premium gas})P(\text{pump premium gas})$$

$$= (0.6)(0.25) = 0.15.$$

From Exercise 12.49, P(pay at least $30) = 0.445. Putting them together,

$$P(\text{pump premium gas} \mid \text{pay at least \$30}) = \frac{0.15}{0.445} = 0.337$$

Exercise 12.57

(a) The possible alleles inherited are B and B, B and O, and O and O. The alleles B and B and B and O both result in a blood type of B for a child. The alleles O and O result in a blood type of O for a child. So the two blood types their children can have are B and O.

(b) Let S_O and S_B correspond to the events that allele O or B is inherited from Sarah, respectively, and D_O and D_B correspond to the events that allele O or B is inherited from David. N_O and N_B each have probability 0.5, and so do D_O and D_B.

$$P(\text{child has type O}) = P(S_O \text{ and } D_O) = 0.5 \times 0.5 = 0.25.$$

You multiply the probabilities because we inherit alleles independently from our mother and father. Since the child must have blood type B or O, the P(child has type B) = $1 - P$(child has type O) = $1 - 0.25$ = 0.75.

CHAPTER 13

BINOMIAL DISTRIBUTIONS

OVERVIEW

One of the most common situations giving rise to a **count X** is the **binomial setting**. The binomial setting consists of four assumptions about how the count was produced:

- The number n of observations is fixed.
- The n observations are all independent.
- Each observation falls into one of two categories called "success" and "failure."
- The probability of success p is the same for each observation.

When these assumptions are satisfied, the number of successes, X, has a **binomial distribution** with n trials and success probability p. For smaller values of n, the probabilities for X can be found easily using statistical software or the exact **binomial probability formula**. The formula is given by

$$P(X = k) = \binom{n}{k} p^k (1 - p)^{n-k}$$

where $k = 0, 1, 2, \cdots, n$, and $\binom{n}{k} = \dfrac{n!}{k!(n-k)!}$ is called the **binomial coefficient**.

When the population is much larger than the sample, a count X of successes in an SRS of size n has approximately the binomial distribution with n equal to the sample size and p equal to the proportion of successes in the population.

The mean of a binomial random variable X is

$$\mu = np$$

and the standard deviation is

$$\sigma = \sqrt{np(1 - p)}$$

When n is large, the count X is approximately $N\left(np, \sqrt{np(1 - p)}\right)$. This approximation should work well when $np \geq 10$ and $n(1 - p) \geq 10$.

GUIDED SOLUTIONS

Exercise 13.1

KEY CONCEPTS: Binomial setting

Four assumptions need to be satisfied to ensure that the count X has a binomial distribution. The number of observations or trials is fixed in advance, each trial results in one of two outcomes, the trials are independent, and the probability of success is the same from trial to trial. In addition, for a large population with a proportion p of successes, we can use the binomial distribution as an approximation to the distribution of the count X of successes in an SRS of size n. For the setting in this exercise, see if all four assumptions are satisfied.

Think about whether random-digit dialing fits in the binomial setting. What is n? What are the two outcomes, and why might the trials be considered independent?

Exercise 13.2

KEY CONCEPTS: Binomial setting

Review the discussion in Exercise 13.1. Think about how many observations or trials there are going to be.

Exercise 13.5

KEY CONCEPTS: Binomial probabilities, binomial tables

(a) Suppose we let X denote the number of errors caught. There are 10 word errors in the essay, and we are in the binomial setting with $n = 10$ trials. Letting "success" correspond to the students catching a word error, we have $p = 0.7$. The distribution of the number of errors caught is $B(10, 0.7)$.

Suppose X denotes the number of errors missed. What are the values of n and p? What is the distribution of the number of errors missed?

(b) X, the number of misses, has the binomial distribution with $n = 10$ and $p = 0.3$. You need to find the probability that $X = 3$. The exact binomial probability formula is given by

$$P(X = k) = \binom{n}{k} p^k (1 - p)^{n-k}$$

and the required probability can be found by plugging the appropriate values of n, k, and p into the formula. Do this to evaluate

$$P(X = 3) =$$

Alternatively, software (or many calculators) can be used to evaluate binomial probabilities such as $P(X = 3)$, although you should do at least one computation by hand to make sure you understand how to use the formula. If you have software, use it to evaluate the probability of missing 3 or more out of 10, or

$$P(X \geq 3) =$$

Exercise 13.9

KEY CONCEPTS: Binomial distribution, mean and variance

(a) Suppose we let X denote the number of errors caught. The distribution of the number of errors caught is $B(10, 0.7)$. The mean of X is

$$\mu = np =$$

Now, suppose X denotes the number of errors missed. The mean of X is

$$\mu = np =$$

What is the total of the number of errors caught plus the number of errors missed? Can you see why the means must add to 10?

(b) Suppose we let X denote the number of errors caught. The distribution of the number of errors caught is $B(10, 0.7)$. The standard deviation of X is

$$\sigma = \sqrt{np(1 - p)} =$$

(c) Suppose $p = 0.9$. The standard deviation of X is

$$\sigma = \sqrt{np(1 - p)} =$$

Now suppose $p = 0.99$. The standard deviation of X is

$$\sigma = \sqrt{np(1 - p)} =$$

What happens to the standard deviation of a binomial distribution as the success probability gets closer to 1?

Exercise 13.11

KEY CONCEPTS: Mean of the binomial, normal approximation for counts

Let X denote the number of home runs Mark McGwire will hit in 509 times at bat. The problem tells us to treat X as a count having the $B(509, 0.116)$ distribution. We are asked to compute the mean μ of X. What is the formula for μ? To answer this, you may want to refer to the Overview for this chapter. Now use this formula to compute the mean.

$$\mu =$$

Compute $P(X \geq 70)$ using the normal approximation to the binomial distribution. First check to see that np and $n(1 - p)$ are both greater than or equal to 10. Next find the mean and standard deviation of X. You have just computed the mean. To compute the standard deviation σ, you may want to refer to the Overview for this chapter.

$$\sigma =$$

Use the normal approximation to evaluate $P(X \geq 70)$. This is a normal probability calculation like those discussed in Chapter 3. You may want to review the material there to refresh your memory on how to do such calculations. The first step is to standardize the number 70 (compute its z-score) by subtracting the mean μ and dividing the result by σ. Next, use Table A to determine the area to the right of this z-score. It may be helpful to draw a normal curve to visualize the area.

$$P(X \geq 70) =$$

Exercise 13.26

KEY CONCEPTS: Binomial probabilities, mean and standard deviation of a binomial count

(a) How many trials are there?   If a success is a rise in the index, what is the success probability?

$$n = \qquad\qquad p =$$

(b) What are the possible values of X?

(c) Use either the exact binomial formula or statistical software to evaluate the probabilities of each possible value of X and then draw a probability histogram for the distribution of X on the next page. You may want to refer to Exercise 13.5 of this Study Guide if you are having difficulties.

$P(X = 0) =$

$P(X = 1) =$

$P(X = 2) =$

$P(X = 3) =$

$P(X = 4) =$

$P(X = 5) =$

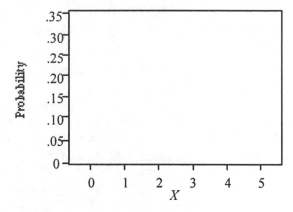

(d) When X is known to have a binomial distribution, you can use the formulas that express the mean and standard deviation of X in terms of n and p. Make sure to mark the location of the mean on your histogram.

Mean =

Standard deviation =

COMPLETE SOLUTIONS

Exercise 13.1

The number of trials, 15, is fixed and each call either succeeds in talking with a live person or it doesn't. Whether or not a particular call succeeds will not alter the probability that any other call succeeds, and the probability of any call succeeding is given as 0.20. The binomial distribution with $n = 15$ and $p = 0.20$ should be a good probability model for the number of calls that reach a live person.

Exercise 13.2

Although each login attempt is a success or a failure, we are not counting the number of successes in a fixed number of attempts. The number of observations (attempts) is random. The assumption of a fixed number of observations is violated.

Exercise 13.5

(a) The distribution of the number of errors caught is given in the Guided Solutions. If X denotes the number of errors missed, then $n = 10$ and $p = 1 - 0.7 = 0.3$, where p is now the probability of missing a given error. The distribution of the number of errors missed is $B(10, 0.3)$.

(b) Letting X denote the number of errors missed, we want $P(X = 3)$. Using the binomial formula to evaluate this probability we have $n = 10$, $k = 3$, and $p = 0.3$, so that

$$P(X = k) = \binom{n}{k} p^k (1 - p)^{n-k} = \binom{10}{3} .3^3 (1 - .3)^{10-3} = \frac{10!}{3!7!} (0.3)^3 (0.7)^7 = 120(0.027)(0.08235) = 0.2668$$

Using software, we have $P(X \geq 3) = 0.6172$.

Exercise 13.9

(a) If X denotes the number of errors caught, the mean of X is $\mu = np = 10(0.7) = 7$. Suppose X denotes the number of errors missed. The mean of X is $\mu = np = 10(0.3) = 3$. We see that these means add to 10. In any experiment, the total of the number of errors caught plus the number of errors missed must *always* be 10, so 10 must be the mean of this total.

(b) If X is the number of errors caught, the standard deviation of X is

$$\sigma = \sqrt{np(1 - p)} = \sqrt{10(0.7)(0.3)} = 1.4491$$

(c) Suppose $p = 0.9$. The standard deviation of X is

$$\sigma = \sqrt{np(1 - p)} = \sqrt{10(0.9)(0.1)} = 0.9487$$

Now suppose $p = 0.99$. The standard deviation of X is

$$\sigma = \sqrt{np(1 - p)} = \sqrt{10(0.99)(0.01)} = 0.3146$$

As the success probability gets closer to 1, the standard deviation gets closer to 0. When the success probability is very close to 1, there is very little variability in the outcome as the proofreader will tend to catch all 10 errors almost every time.

Exercise 13.11

(a) $\mu = np = 509 \times 0.116 = 59.044$

(b) First we check that

$$np = 509 \times 0.116 = 59.044 \geq 10 \text{ and } n(1 - p) = 509 \times 0.884 = 449.956 \geq 10$$

Next we compute the mean and standard deviation of X.

$$\mu = np = 509 \times 0.116 = 59.044$$

$$\sigma = \sqrt{np(1-p)} = \sqrt{509 \times 0.116 \times 0.884} = \sqrt{52.195} = 7.225$$

When n is large, X is approximately $N\left(np, \sqrt{np(1-p)}\right) = N(59.044, 7.225)$. Thus

$$z\text{-score of } 70 = \frac{70 - 59.044}{7.225} = 1.52$$

and using Table A,

$$P(X \geq 70) = P(Z \geq 1.52) = 1 - P(Z \leq 1.52) = 1 - 0.9357 = 0.0643$$

Exercise 13.26

(a) X has a binomial distribution with $n = 5$ (the number of years to be observed) and $p = 0.65$ (the probability the index will increase in any given year). The independence of years is assumed as part of the model.

(b) Because $n = 5$, the possible values are X are 0, 1, 2, 3, 4, 5.

(c) To calculate the probability of each value of X, we can use the binomial formula or statistical software. This exercise is very similar to Exercise 12.5 of this Study Guide in which the use of the binomial formula was illustrated. The only difference is that $p = 0.65$ in this exercise and p was 0.25 in Exercise 12.5. The probabilities listed were obtained using the Minitab software.

```
Binomial with n = 5 and p = 0.650000
     x          P(X = x)
   0.00          0.0053
   1.00          0.0488
   2.00          0.1811
   3.00          0.3364
   4.00          0.3124
   5.00          0.1160
```

The probability histogram corresponding to this distribution follows.

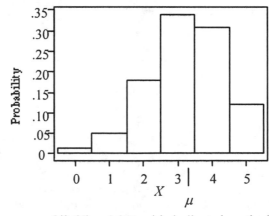

(d) The mean of X is $\mu = np = 5(0.65) = 3.25$ and is indicated on the histogram in part (c). The standard deviation of X is

$$\sigma = \sqrt{np(1-p)} = \sqrt{5(0.65)(0.35)} = 1.067$$

CHAPTER 14

CONFIDENCE INTERVALS: THE BASICS

OVERVIEW

A **confidence interval** provides an estimate of an unknown parameter of a population or process, along with an indication of how accurate this estimate is and how confident we are that the interval is correct. Confidence intervals have two parts. One is an interval computed from our data, typically of the form

$$\text{estimate} \pm \text{margin of error}$$

The other part is the **confidence level,** which states the probability that the *method* used to construct the interval will give a correct answer. For example, if you use a 95% confidence interval repeatedly, in the long run 95% of the intervals you construct will contain the correct parameter value. Of course, when you apply the method only once, you do not know whether your interval gives a correct value or not. Confidence refers to the probability that the method gives a correct answer in repeated use, not the correctness of any particular interval we compute from data.

Suppose we wish to estimate the unknown mean μ of a normal population with known standard deviation σ based on an SRS of size n. A level C confidence interval for μ is

$$\bar{x} \pm z^* \frac{\sigma}{\sqrt{n}}$$

where z^* is such that the probability is C that a standard normal random variable lies between $-z^*$ and z^* and is obtained from the bottom row in Table C. These z-values are called critical values.

The margin of error $z^* \frac{\sigma}{\sqrt{n}}$ of a confidence interval decreases when any of the following occur:

- The confidence level C decreases.

- The sample size n increases.

- The population standard deviation σ decreases.

The sample size needed to obtain a confidence interval for a normal mean of the form

$$\text{estimate} \pm \text{margin of error}$$

with a specified margin of error m is

$$n = \left(\frac{z^* \sigma}{m} \right)^2$$

where z^* is the critical value for the desired level of confidence. Many times, the n you find is not be an integer. If it is not, round up to the next larger integer.

The formula for any specific confidence interval is a recipe that is correct under specific conditions. The most important conditions concern the methods used to produce the data. Many methods (including those discussed here) assume that the data were collected by random sampling. Other conditions, such as the actual distribution of the population, are also important.

GUIDED SOLUTIONS

Exercise 14.2

KEY CONCEPTS: Confidence intervals, interpreting statistical confidence

(a) The general form of a confidence interval is

estimate ± margin of error

Identify the estimate (the percentage of people who said yes when asked, "Would you like to lose weight?") and the margin of error. Then combine the estimate and the margin of error in the general form of a confidence interval as indicated.

(b) In formulating your explanation, consider the meaning of statistical confidence as described in Chapter 14 of your textbook or in the Overview for this chapter of the Study Guide. Write your explanation.

Exercise 14.7

KEY CONCEPTS: Confidence intervals for means, the effect of sample size on the margin of error

(a) Identify the following quantities. You will want to use the bottom row of Table C to find z^*.

$\bar{x} =$

$\sigma =$

$n =$

z^* (for a 95% confidence interval) =

Use the formula for a 95% confidence interval to compute the desired interval.

$$\bar{x} \pm z^* \frac{\sigma}{\sqrt{n}} =$$

(b) Use the same formula as in part (a) but now with $n = 250$ rather than 1000.

$$\bar{x} \pm z^* \frac{\sigma}{\sqrt{n}} =$$

(c) Use the same formula as in part (a) but now with $n = 4000$.

$$\bar{x} \pm z^* \frac{\sigma}{\sqrt{n}} =$$

(d) The margins of errors are the quantities after the $\pm$. You computed them in parts (a), (b), and (c). List them here.

Margin of error for $n = 250$:

Margin of error for $n = 1000$:

Margin of error for $n = 4000$:

What pattern do you observe?

Exercise 14.9

KEY CONCEPTS: Sample size required to obtain a confidence interval of specified margin of error

Refer to Exercise 14.7, part (a), of this Study Guide for the appropriate values of σ and z^*. What margin of error m is desired? Complete the following to compute the necessary sample size n.

$$n = \left(\frac{z^* \sigma}{m} \right)^2 =$$

Exercise 14.25

KEY CONCEPTS: Confidence intervals for means, the effect of changing the confidence level

(a) Make the stemplot with the stems in the space provided. Notice that the smallest value in our data is –8.3 and the largest is 2.2, so the stems range from –8 to 2 with both a 0 and –0 stem. Refer to Chapter 1 of your textbook if you need to refresh your memory about stemplots.

```
-8 |
-7 |
-6 |
-5 |
-4 |
-3 |
-2 |
-1 |
-0 |
 0 |
 1 |
 2 |
```

Are there any outliers? Is there extreme skewness? Do the data appear to follow a normal distribution closely?

(b) The four-step process follows.

State. What is the practical question in the context of the real-world setting?

Formulate. What specific statistical operations does this problem call for?

Solve. Make the graphs and carry out any calculations needed for this problem.

Conclude. Give your practical conclusion in the setting of the real-world problem.

To apply the steps to this exercise, here are some suggestions.

State. What characteristic of bone mineral content of breast-feeding mothers is of interest here?

Formulate. What does the problem ask for?

Solve. Are the conditions for inference satisfied? (Do we have an SRS? Is the population approximately Normal? Do we know σ?) What does part (a) tell you? Now identify $\sigma, n, \bar{x}$ (you will need to calculate them), and z^* for a 99% confidence interval (use Table C). Then use the formula to calculate the confidence interval.

$$\bar{x} \pm z^* \frac{\sigma}{\sqrt{n}} =$$

Conclude. State clearly what you have found in terms of the change in the mean bone mineral content of the population of all breast-feeding mothers.

Exercise 14.35

KEY CONCEPTS: Interpreting confidence intervals

Review the meaning of statistical confidence. In the long run, what is supposed to be within 3 percentage points of at least 95% of the results?

Exercise 14.36

KEY CONCEPTS: Effect of sample size on the margin of error

Is the number of respondents aged 18 to 29 years larger than, smaller than, or equal to 1002? What effect does a change in sample size have on the margin of error?

COMPLETE SOLUTIONS

Exercise 14.2

(a) The estimate of interest here (the percent of people who said yes when asked, "Would you like to lose weight?") is 51%. Since the margin of error for a 95% confidence interval is ± 3%, the 95% confidence interval for the percent of all adult women who claim they would like to lose weight is 51% ± 3%, or between 48% and 54%.

(b) Suppose we take all possible random samples of national adults. In each sample, suppose we determine the percent of people who say yes when asked, "Would you like to lose weight?" For each of these percents, suppose we add and subtract the margin of error for a 95% confidence interval. Of the resulting intervals, 95% will contain the actual percent of all adults in the United States who would like to lose weight. This is what we mean by "95% confidence." Note that we do not know if any particular interval (such as the 48% to 54% interval in part (a)) contains the true value of the percent. The confidence level of 95% refers only to the percent of the intervals produced by all samples that will contain the true percent.

Exercise 14.7

(a) We are given that the sample mean $\bar{x}$ is 22, the standard deviation σ is 50, and the sample size n is 1000. For 95% confidence, $z^* = 1.96$. Thus a 95% confidence interval for the mean gain in score μ in the population of all high school students is

$$\bar{x} \pm z^* \frac{\sigma}{\sqrt{n}} = 22 \pm 1.96 \frac{50}{\sqrt{1000}} = 22 \pm 3.10$$

or 18.90 to 25.10.

(b) We simply replace $n = 1000$ by $n = 250$ in our calculations:

$$\bar{x} \pm z^* \frac{\sigma}{\sqrt{n}} = 22 \pm 1.96 \frac{50}{\sqrt{250}} = 22 \pm 6.2, \text{ or } 15.80 \text{ to } 28.20$$

(c) We use $n = 4000$ in our calculations:

$$\bar{x} \pm z^* \frac{\sigma}{\sqrt{n}} = 22 \pm 1.96 \frac{50}{\sqrt{4000}} = 22 \pm 1.55, \text{ or } 19.45 \text{ to } 23.55$$

(d)   Margin of error for $n = 250$: ± 6.20

Margin of error for $n = 1000$: ± 3.10

Margin of error for $n = 4000$: ± 1.55

We see that as the sample size increases, the margin of error decreases.

Exercise 14.9

We want a 95% confidence interval with a margin of error $m = 2$. As we saw in part (a) of Exercise 14.7 of this Study Guide, $\sigma = 50$, and from Table C we find $z^* = 1.96$ for a 95% confidence interval. The formula for the proper sample size n is

$$n = \left(\frac{z^* \sigma}{m}\right)^2 = \left(\frac{1.96 \times 50}{2}\right)^2 = 2401$$

which is (conveniently) an integer.

Exercise 14.25

(a) Here is a stemplot of the data. We have used split stems. There are no outliers. There does seem to be very slight right skewness, but it is certainly not extreme. Overall the stemplot follows a normal distribution quite well.

```
-8 | 3
-7 | 08
-6 | 25588
-5 | 12233679
-4 | 0347799
-3 | 01368
-2 | 011223557
-1 | 008
-0 | 38
 0 | 234
 1 | 7
 2 | 2
```

(b) *State*. The problem tells us that we are interested in estimating the mean percent change during three months of breast-feeding in the bone mineral content of the spines of the population of all breast-feeding mothers.

Formulate. The problem asks us to use a 99% confidence interval to estimate this mean percent change in the population.

Solve. In part (a) we are told that the researchers were willing to consider these 47 women as an SRS from the population of all nursing mothers (of course, this assumption and actually selecting them by simple random sampling are not the same). In part (a) we also saw that the distribution of the percent change in bone mineral content appeared to follow a normal distribution. Finally, we know σ. Thus, the conditions for inference appear to be satisfied.

We are told that $\sigma = 2.5$ and we know that $n = 47$. From the data we calculate $\bar{x} = -3.587$. From Table C we find $z^* = 2.576$ for a 99% confidence interval. Thus our 99% confidence interval is

$$\bar{x} \pm z^* \frac{\sigma}{\sqrt{n}} = -3.587 \pm 2.576 \frac{2.5}{\sqrt{47}} = -3.587 \pm 0.939, \text{ or } (-4.526, -2.648)$$

Conclude. We are 99% confident that the mean percent change during three months of breast-feeding in the bone-mineral content of the spines of the population of all breast-feeding mothers is between −4.526 and −2.648.

Exercise 14.35

If you use a 95% confidence interval repeatedly, in the long run 95% of the intervals you construct will contain the correct *parameter value* (in this case the percentage of all adults), not the result of this particular survey. Thus, the last sentence in the Associated Press quote should be "This means that, if the same questions were repeated in 20 polls, the results of at least 19 surveys would be within three percentage points of the percentage of *all adults* in the population."

Exercise 14.36

The poll consists of 1002 adults. Respondents aged 18 to 29 years are only a portion of all respondents, and the number of respondents aged 18 to 29 years must be less than 1002. Recall that the margin of error increases as sample size increases. Thus, the margin of error for respondents aged 18 to 29 years must be larger than ± 3 percentage points.

CHAPTER 15

TESTS OF SIGNIFICANCE: THE BASICS

OVERVIEW

Tests of significance and confidence intervals are the two most widely used types of formal statistical inference. A test of significance is done to assess the evidence against the **null hypothesis** H_0 in favor of an **alternative hypothesis** H_a. Typically, the alternative hypothesis is the effect that the researcher is trying to demonstrate, and the null hypothesis is a statement that the effect is not present. The alternative hypothesis can be either **one-** or **two-sided**.

Tests are usually carried out by first computing a **test statistic**. The test statistic is used to compute a **P-value**, which is the probability of getting a test statistic at least as extreme as the one observed, where the probability is computed when the null hypothesis is true. The P-value provides a measure of how incompatible our data are with the null hypothesis, or how unusual it would be to get data like ours if the null hypothesis were true. Since small P-values indicate data that are unusual or difficult to explain under the null hypothesis, we typically reject the null hypothesis in these cases. In this case, the alternative hypothesis provides a better explanation for our data.

Significance tests of the null hypothesis H_0: $\mu = \mu_0$ with either a one- or a two-sided alternative are based on the test statistic

$$z = \frac{\bar{x} - \mu_0}{\sigma / \sqrt{n}}$$

The use of this test statistic assumes that we have an SRS from a normal population with known standard deviation σ. When the sample size is large, the assumption of normality is less critical because the sampling distribution of $\bar{x}$ is approximately normal. P-values for the test based on z are computed using Table A.

When the P-value is less than a specified value α, we say that the results are **statistically significant at level α,** or we reject the null hypothesis at level α. Tests can be carried out at a fixed significance level by obtaining the appropriate critical value z^* from the bottom row in Table C.

GUIDED SOLUTIONS

Exercise 15.2

KEY CONCEPTS: Testing hypotheses about means

(a) If the null hypothesis H_0: $\mu = 115$ is true, then scores in the population of older students are normally distributed, with mean $\mu = 115$ and standard deviation $\sigma = 30$. What then is the sampling distribution of $\bar{x}$, the mean of a sample of size $n = 25$? (We studied the sampling distribution of $\bar{x}$ in Chapter 11.)

Sketch the density curve of this distribution. Be sure to label the horizontal axis properly.

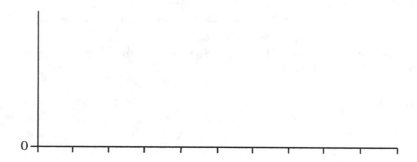

(b) Mark the two points on your sketch in part (a). Referring to the sketch, explain in simple language why one result is good evidence that the mean score of all older students is greater than 115 and why the other outcome is not. Think about how far out on the density curve the two points are.

Exercise 15.10

KEY CONCEPTS: Testing hypotheses about means, the z-test statistic

We are testing the hypothesis H_0: $\mu = 115$, so $\mu_0 = 115$. The two outcomes in Exercise 15.2 are $\bar{x} = 118.6$ and $\bar{x} = 125.8$. The population standard deviation is $\sigma = 30$, and the sample size is $n = 25$. To compute the values of the test statistic z, complete the following.

For $\bar{x} = 118.6$: $z = \dfrac{\bar{x} - \mu_0}{\sigma/\sqrt{n}} =$

For $\bar{x} = 125.8$: $z = \dfrac{\bar{x} - \mu_0}{\sigma/\sqrt{n}} =$

Exercise 15.14

KEY CONCEPTS: *P*-values, statistical significance

Refer to the graph that follows. The *P*-value for 118.6 is the shaded area under the normal curve to the right of 118.6. We learned how to calculate such areas in Chapter 3. First you will need to find the *z* score of 118.6 and then you will need to use Table A to find the area to the right of this *z*-score under a standard normal curve. Use the space for your calculations.

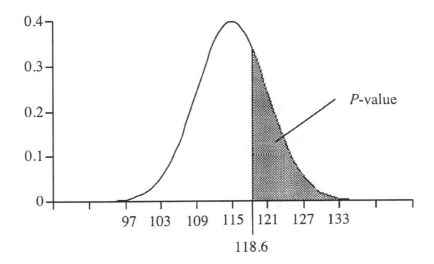

Now perform a similar calculation to find the *P*-value of 125.8.

Why do these values tell us that one outcome is strong evidence against the null hypothesis and that the other is not?

Exercise 15.16

KEY CONCEPTS: *P*-values, statistical significance

To answer this question, recall that an observed value is statistically significant at level α if the *P*-value is smaller than α. Use this fact and your solutions to Exercise 15.14 to answer the question.

Exercise 15.23

KEY CONCEPTS: Testing hypotheses at a fixed significance level

(a) The *z*-test statistic for testing against a two-sided alternative, as in this problem, is $|z| = \left|\dfrac{\bar{x} - \mu_0}{\sigma/\sqrt{n}}\right|$. Identify μ_0, the standard deviation σ, the sample mean $\bar{x}$, and the sample size n. Then complete the test.

$$|z| = \left|\frac{\bar{x} - \mu_0}{\sigma/\sqrt{n}}\right| =$$

(b) To answer, you will have to find the appropriate critical value from Table C. Note that we are testing against a two-sided alternative.

(c) Follow the procedure in part (b) but with significance level 1%.

(d) Between what two adjacent critical values in Table C does your *z*-test statistic lie? What are the corresponding tail probabilities at the top of the table? What do you need to do to the tail probabilities to convert them to critical values for testing against a two-sided alternative?

Exercise 15.25

KEY CONCEPTS: Relationship between two-sided tests and confidence intervals

(a) The 95% confidence interval 31.5 ± 3.5 is equivalent to the interval 28 to 35. What must be true of the relationship between a $1 - \alpha$ confidence interval for μ and the value μ_0 for a level α two-sided significance test of H_0: $\mu = \mu_0$ to reject the null hypothesis? Is it true here?

(b) Is the condition mentioned in part (a) true here?

Exercise 15.39

KEY CONCEPTS: Testing hypotheses about means

The four-step process follows.

State. What is the practical question, in the context of the real-world setting?

Formulate. What specific statistical operations does this problem call for?

Solve. Make the graphs and carry out any calculations needed for this problem.

Conclude. Give your practical conclusion in the setting of the real-world problem.

To apply the steps to this problem, here are some suggestions.

State. What characteristic of bone mineral content of breast-feeding mothers is of interest here? What question about this characteristic do we wish to answer?

Formulate. What hypotheses do we need to test? Is the alternative one-sided or two-sided?

Solve. Are the conditions for inference satisfied? (Do we have an SRS? Is the population approximately normal? Do we know σ?) You may need to refer to Exercise 14.25. Identify σ, n, and $\bar{x}$ (you will need to calculate $\bar{x}$ and may have already done so if you solved Exercise 14.25) and calculate z.

$$z = \frac{\bar{x} - \mu_0}{\sigma / \sqrt{n}} =$$

Now compute the *P*-value.

Conclude. State clearly what you have found in terms of the change in the mean bone mineral content of the population of all breast-feeding mothers.

Exercise 15.52

KEY CONCEPTS: Interpreting *P*-values

Recall that the *P*-value is the probability, computed supposing H_0 to be true, that the test statistic will take a value at least as extreme as that actually observed. Use this to evaluate the student's statement.

COMPLETE SOLUTIONS

Exercise 15.2

(a) From Chapter 10 we know that the sampling distribution of $\bar{x}$ is normal with mean $\mu = 115$ and standard deviation $\sigma = 30/\sqrt{n} = 30/\sqrt{25} = 30/5 = 6$. A sketch of the density curve of this distribution follows.

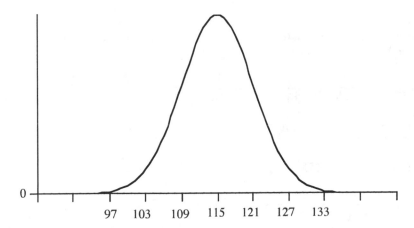

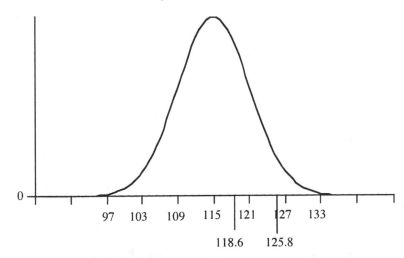

(b) The two points are marked on the following curve.

The 125.8 is much farther out on the normal curve than 118.6. In other words, it would be unlikely to observe a mean of 125.8 if the null hypothesis H_0: $\mu = 115$ is true. However, a mean of 118.6 is fairly likely if the null hypothesis is true. A mean as large as 125.8 is more likely to occur if the true mean is larger than 115. Thus, 125.8 is good evidence that the mean score of all older students is greater than 115, while a mean score of 118.6 is not.

Exercise 15.10

For $\bar{x} = 118.6$: $z = \dfrac{\bar{x} - \mu_0}{\sigma / \sqrt{n}} = \dfrac{118.6 - 115}{30 / \sqrt{25}} = \dfrac{3.6}{6} = 0.6$

For $\bar{x} = 125.8$: $z = \dfrac{\bar{x} - \mu_0}{\sigma / \sqrt{n}} = \dfrac{125.8 - 115}{30 / \sqrt{25}} = \dfrac{10.8}{6} = 1.8$

Exercise 15.14

Refer to the graph in the guided solution. The P-value for 118.6 is the shaded area under the normal curve to the right of 118.6. We learned how to calculate such areas in Chapter 3. First we compute the z-score of 118.6. We did this in Exercise 14.10 and found

$$z\text{-score} = \frac{\bar{x} - \mu_0}{\sigma/\sqrt{n}} = \frac{118.6 - 115}{30/\sqrt{25}} = \frac{3.6}{6} = 0.6$$

From Table A we find that the area under the standard normal curve to the right of 0.6 is 1 minus the area to the left of 0.6 = $1 - 0.7257 = 0.2743$.

We make a similar calculation for the P-value of 125.8.

$$z\text{ score} = \frac{\bar{x} - \mu_0}{\sigma/\sqrt{n}} = \frac{125.8 - 115}{30/\sqrt{25}} = \frac{10.8}{6} = 1.8$$

The area to the right of this z-score under a standard normal curve = 1 minus the area to the left of 1.80 = $1 - 0.9641 = 0.0359$.

In summary,

$$P\text{-value of } 118.6 = 0.2743$$

$$P\text{-value of } 125.8 = 0.0359$$

Using software we find that the P-value of 118.6 is 0.27425312 and the P-value of 0.0359 is 0.03593032.

The P-value for 118.6 is not particularly small, so this outcome is reasonably likely to happen by chance if the null hypothesis is true. Thus, the outcome 118.6 is not strong evidence against the null hypothesis. The P-value for 125.8 is small, so this outcome is unlikely to happen by chance if the null hypothesis is true. Thus, the outcome 125.8 is strong evidence against the null hypothesis.

Exercise 15.16

The P-value for 125.8 is less than 0.05, so it is statistically significant at the $\alpha = 0.05$ level. The P-value for 118.6 is larger than 0.05, so it is not statistically significant at the 0.05 level.

Neither P-value is less than 0.01, so neither observed value would be statistically significant at $\alpha = 0.01$.

Exercise 15.23

(a) Because the null hypothesis is H_0: $\mu = 0.5$, we have $\mu_0 = 0.5$. The standard deviation is $\sigma = 0.2887$, the sample mean is $\bar{x} = 0.4365$, and the sample size is $n = 100$; the value of the z-test statistic is

$$|z| = \left| \frac{\bar{x} - \mu_0}{\sigma/\sqrt{n}} \right| = \left| \frac{0.4365 - 0.5}{0.2887/\sqrt{100}} \right| = \left| \frac{-0.0635}{0.02887} \right| = |{-2.20}| = 2.20$$

(b) Because we are testing against a two-sided alternative, we need to find the upper $\alpha/2 = 0.05/2 = 0.025$ critical value in Table C. We see that this critical value is 1.96. Since our z-test statistic is greater than this critical value, the result is significant at the 0.05 level.

(c) Now we look for the upper $\alpha/2 = 0.01/2 = 0.005$ critical value in Table C. We see that this critical value is 2.576. Since our z-test statistic is smaller than this critical value, the result is not significant at the 0.01 level.

(d) As we examine the critical values in Table C, we observe that our z-test statistic of 2.20 lies between the critical values 2.054 and 2.326. The corresponding upper tail probabilities at the top of the table are 0.02 and 0.01. Since we are testing against a two-sided alternative, we must multiply the tail probabilities by 2 to get the appropriate critical values (we double the values because we need to include both the upper and lower tail probabilities). Therefore we see that 2.20 lies between the 0.04 and 0.02. We conclude that $0.02 < P$-value < 0.04.

Statistical software tells us that the P-value is 0.0278069.

Exercise 15.25

(a) The null hypothesis is H_0: $\mu = 34$. The value of μ under H_0, namely 34, falls inside the 95% confidence interval and we would not reject H_0.

(b) The null hypothesis is H_0: $\mu = 36$. The value of μ under H_0, namely 36, falls outside the 95% confidence interval and we would reject H_0.

Exercise 15.39

State. Exercise 4.25 and this exercise tells us that we are interested in the mean percent change during three months of breast-feeding in the bone mineral content of the spines of the population of all breast-feeding mothers. The problem further tells us that we want to know if the data give good evidence that on the average nursing mothers lose bone mineral. A loss of bone mineral would correspond to a mean percent change that is less than 0.

Formulate. Let μ denote the mean percent change in the mineral content of the spine in the population of all nursing mothers. The alternative is one-sided (because we are interested in whether the mean percent change is less than 0), and we want to test the hypotheses

$$H_0: \mu = 0$$

$$H_a: \mu < 0$$

Solve. In Exercise 4.25 we are told that the researchers were willing to consider these 47 women as an SRS from the population of all nursing mothers (of course, this assumption and actually selecting them by simple random sampling are not the same). The problem tells us that the distribution of the percent change in bone mineral content in the population appears to follow a normal distribution with standard deviation $\sigma = 2.5$. Thus, the conditions for inference appear to be satisfied.

We know that $n = 47$. From the data, we calculate $\bar{x} = -3.587$. Thus,

$$z = \frac{\bar{x} - \mu_0}{\sigma/\sqrt{n}} = \frac{-3.587 - 0}{2.5/\sqrt{47}} = -9.836.$$

The alternative is one-sided, so the P-value in this case is the area under the standard normal curve to the left of -9.836, and -9.836 is outside the range of values given in Table A. All that we can say is that

$$P\text{-value} < 0.0002$$

Software gives the exact value of 3.9391×10^{-23}.

Conclude. The P-value is so small that we conclude the data give strong evidence that the mean percent change during three months of breast-feeding in the bone-mineral content of the spines of the population of all breast-feeding mothers is less than 0.

Exercise 15.52

The student's statement is not correct. The null hypothesis is either true or false; statements about the probability that it is true are not meaningful. The P-value is the probability, computed supposing H_0 to be true, that the test statistic will take a value at least as extreme as that actually observed. Thus, P-values tell us about how strong our data are as evidence against the null hypotheses. Small values indicate strong evidence against the null hypothesis.

CHAPTER 16

INFERENCE IN PRACTICE

OVERVIEW

Statistical inference from data based on a badly designed survey or experiment is often useless. Remember, a statistical test is valid only under certain conditions with data that have been properly produced. Whenever you use statistical inference, you are assuming that your data are a probability sample or come from a randomized comparative experiment.

Always do data analysis before inference to detect outliers or other problems that would make your inference untrustworthy.

The margin of error of a confidence interval accounts only for the chance variation due to random sampling. Errors due to nonresponse or undercoverage are often more serious in practice.

When describing the outcome of a hypothesis test, it is more informative to give the P-value than to just reject or not reject a decision at a particular significance level α. The traditional levels of 0.01, 0.05, and 0.10 are arbitrary and serve as rough guidelines. Different people insist on different levels of significance depending on the plausibility of the null hypothesis and the consequences of rejecting the null hypothesis. There is no sharp boundary between significant and insignificant, only increasingly strong evidence as the P-value decreases.

When testing hypotheses with a very large sample, the P-value can be very small for effects that may not be of interest. Don't confuse small P-values with large or important effects. Statistical significance is not the same as practical significance. Plot the data to display the effect you are trying to show and also give a confidence interval that says something about the size of the effect.

Just because a test is not statistically significant doesn't imply that the null hypothesis is true. Statistical significance may occur when the test is based on a small sample size. Finally, if you run enough tests, you will invariably find statistical significance for one of them. Be careful in interpreting the results when testing many hypotheses on the same data.

From the point of view of making decisions, H_0 and H_a are just two statements of equal status that we must decide between. One chooses a rule for deciding between H_0 and H_a on the basis of the probabilities of the two types of errors we can make. A **Type I error** occurs if H_0 is rejected when it is in fact true. A **Type II error** occurs if H_0 is accepted when in fact H_a is true. There is a clear relation between α-level significance tests and testing from the decision-making point of view. The probability of a Type I error is α.

To compute the Type II error probability of a significance test about a mean of a normal population:

   • Write the rule for accepting the null hypothesis in terms of $\bar{x}$.

   • Calculate the probability of accepting the null hypothesis when the alternative is true.

The **power** of a significance test is always calculated at a specific alternative hypothesis and is the probability that the test will reject H_0 when that alternative is true. The power of a test against any particular alternative is 1 minus the probability of a Type II error. Power is usually interpreted as the ability of a test to detect an alternative hypothesis or as the sensitivity of a test to an alternative hypothesis. The power of a test can be increased by increasing the sample size when the significance level remains fixed.

GUIDED SOLUTIONS

Exercise 16.3

KEY CONCEPTS: Sources of error and confidence intervals

To answer these questions, recall that the margin of error of a confidence interval accounts only for the chance variation due to random sampling. Errors due to nonresponse or undercoverage are often more serious in practice.

(a) Is this source of error included in the margin of error? Yes _____   No _____

(b) Is this source of error included in the margin of error? Yes _____   No _____

(c) Is this source of error included in the margin of error? Yes _____   No _____

Exercise 16.7

KEY CONCEPTS: Statistical significance and practical importance

In this exercise we see that the P-value associated with the outcome $\bar{x} = 4.8$ depends on the sample size. The probability of getting a value of $\bar{x}$ as small as 4.8 if the true mean is 5 becomes smaller as the sample size gets larger. (Do you remember why? Look at the formula for the z score of $\bar{x}$.) Since this probability is the P-value, we see that a small effect is more likely to be detected for larger sample sizes than for smaller sample sizes. But this doesn't necessarily make the effect interesting or important. A confidence interval tells you something about the size of the effect, not the P-value.

(a) Find the P-value by computing the z-test statistic and the probability of being less that the z-test statistic.

(b) This exercise is the same as that in part (a) but with a larger sample size. The larger sample size makes the probability of getting a value of $\bar{x}$ as small as 4.8 smaller than it was in part (a). Compute the *P*-value.

(c) Find the *P*-value in this case. It will be the smallest. Why?

Exercise 16.10

KEY CONCEPTS: Multiple analyses

(a) What does a *P*-value less than 0.01 mean? Out of 500 subjects, how many would you expect to achieve a score that has such a *P*-value if all 500 are guessing?

(b) What would you suggest the researcher do now to test whether any of these four subjects have ESP?

Exercise 16.11

KEY CONCEPTS: Sample size

(a) We want a power of 90% for a 5% level of significance. The small table in the text that follows the definition of power gives the required sample sizes. To use the table, we need to calculate the effect size corresponding to the specified true mean. We know that the hypothesized mean is 115 and the standard deviation is assumed to be 30. Compute the effect size for a true mean of 130.

$$\text{effect size} = \frac{\text{true mean response} - \text{hypothesized mean response}}{\text{standard deviation}} =$$

What does the table give as the required sample size?

(b) Here the true mean is 139.

$$\text{effect size} = \frac{\text{true mean response} - \text{hypothesized mean response}}{\text{standard deviation}} =$$

What does the table give as the required sample size?

Exercise 16.15

KEY CONCEPTS: Type I and Type II error probabilities

(a) Write the two hypotheses. Remember, we usually take the null hypothesis to be the statement of "no effect."

   H_0:

   H_a:

Describe the two types of errors as "false positive" and "false negative" test results.

(b) Which error probability would you choose to make smaller (at the expense of making the other error probability larger) and why?

Exercise 16.18

KEY CONCEPTS: Type I and Type II error probabilities

(a) If $\mu = 0$, what is the sampling distribution of $\bar{x}$? Use the sampling distribution to compute the probability that the test rejects — the probability $\bar{x} > 0$.

(b) If $\mu = 0.3$, what is the sampling distribution of $\bar{x}$? Use the sampling distribution to compute the probability that the test accepts H_0 — the probability $\bar{x} \leq 0$.

(c) If $\mu = 1$, what is the sampling distribution of $\bar{x}$? Use the sampling distribution to compute the probability that the test accepts H_0 — the probability $\bar{x} \leq 0$.

Exercise 16.48

KEY CONCEPTS: Power and its relationship with the Type II error probability

What is the relationship between the probability of a Type I error and the level of significance?

What is the relationship between the power of a test at a particular alternative and the Type II error at this alternative? Now use the fact that the value of the power is 0.78 to compute the Type II error probability.

COMPLETE SOLUTIONS

Exercise 16.3

(a) Errors due to undercoverage (as is the case here) are *not* included in the margin of error.

(b) Errors due to nonresponse (as is the case here) are *not* included in the margin of error.

(c) The margin of error of a confidence interval accounts only for the chance variation due to random sampling. So the error here *is* included in the margin of error.

Exercise 16.7

See the guided solutions for a full explanation of the way sample size can change *P*-values.

(a) The *z*-test statistic is

$$z = \frac{\bar{x} - \mu_0}{\sigma/\sqrt{n}} = \frac{4.8 - 5}{0.5/\sqrt{5}} = -0.89$$

and

$$P\text{-value} = P(Z < -0.89) = 0.1867$$

because the alternative is one-sided. Note that a more exact calculation using statistical software uses $z = -0.8944$ and gives a *P*-value of 0.18555396.

(b) The test statistic is

$$z = \frac{\bar{x} - \mu_0}{\sigma/\sqrt{n}} = \frac{4.8 - 5}{0.5/\sqrt{15}} = -1.55$$

and

$$P\text{-value} = P(Z < -1.55) = 0.0606$$

Note that a more exact calculation using statistical software uses $z = -1.5492$ and gives a *P*-value of 0.06066682.

(c) The test statistic is

$$z = \frac{\bar{x} - \mu_0}{\sigma/\sqrt{n}} = \frac{4.8 - 5}{0.5/\sqrt{40}} = -2.53$$

and

$$P\text{-value} = P(Z < -2.53) = 0.0057$$

Note that a more exact calculation using statistical software uses $z = -2.5298$ and gives a *P*-value of 0.00570638.

Exercise 16.10

(a) A P-value of 0.01 means that the probability a subject would do so well when merely guessing is only 0.01. Among 500 subjects, all of whom are merely guessing, we would therefore expect 1%, or 5, of them to do significantly better than random guessing ($P < 0.01$). Thus in 500 tests it is not unusual to see four results with P-values on the order of 0.01, even if all are guessing and none have ESP.

(b) These four subjects only should be retested with a new, well-designed test. If all four again have low P-values (say, below 0.01 or 0.05), we have real evidence that they are not merely guessing. In fact, if any one of the subjects has a very low P-value (say, below 0.01), it would also be reasonably compelling evidence that the individual is not merely guessing. A single P-value on the order of 0.10, however, would not be particularly convincing.

Exercise 16.11

(a) effect size $= \dfrac{\text{true mean response } - \text{ hypothesized mean response}}{\text{standard deviation}} = \dfrac{130 - 115}{30} = 0.5$

According to the small table in the text after the definition of power, a sample size of 35 is needed.

(b) effect size $= \dfrac{\text{true mean response } - \text{ hypothesized mean response}}{\text{standard deviation}} = \dfrac{139 - 115}{30} = 0.8$

According to the small table in the text after the definition of power, a sample size of 14 is needed.

Exercise 16.15

(a) The two hypotheses are

H_0: the patient has no medical problem
H_a: the patient has a medical problem

One possible error is to decide

H_a: the patient has a medical problem

when, in fact, the patient does not really have a medical problem. This is a Type I error and in this setting could be called a false positive. The other type of error is to decide

H_0: the patient has no medical problem

when, in fact, the patient does have a problem. This is a Type II error and in this setting could be called a false negative.

(b) Most probably we would choose to decrease the error probability for a Type II error or for the false negative probability. Failure to detect a problem (particularly a major problem) when one is present could result in serious consequences (such as death). Although a false positive can also have serious consequences (painful or expensive treatment that is not necessary), it is not likely to lead to the kinds of consequences that a false negative could produce. For example, consider the consequences of failure to detect a heart attack, the presence of AIDS, or the presence of cancer. Note, that there are cases where some might argue that a false positive would be a more serious error than a false negative. For example, a false positive in a test for Down's syndrome or a birth defect in an unborn baby might lead parents to consider an abortion. Some would consider this a much more serious error than to give birth to a child with a birth defect.

Exercise 16.18

(a) If $\mu = 0$, the sampling distribution of $\bar{x}$ is normal with mean $\mu = 0$ and standard deviation $\frac{\sigma}{\sqrt{n}} = \frac{1}{\sqrt{9}} = 0.33$. Thus the probability of a Type I error is the probability that $\bar{x} > 0$ when the null hypothesis is true. Computing the z score for $\bar{x}$, we get

$$P(\bar{x} > 0) = P\left(\frac{\bar{x} - 0}{0.33} > \frac{0 - 0}{0.33} \right) = P(Z > 0) = 0.5$$

(b) If $\mu = 0.3$, the sampling distribution of $\bar{x}$ is normal with mean $\mu = 0.3$ and standard deviation $\frac{\sigma}{\sqrt{n}} = \frac{1}{\sqrt{9}} = 0.33$. We accept H_0 if $\bar{x} \leq 0$. Thus the probability of a Type II error when $\mu = 0.3$ is

$$P(\bar{x} > 0) = P\left(\frac{\bar{x} - 0.3}{0.33} \leq \frac{0 - 0.3}{0.33} \right) = P(Z \leq -0.91) = .1814$$

(c) If $\mu = 1$, the sampling distribution of $\bar{x}$ is normal with mean $\mu = 1$ and standard deviation $\frac{\sigma}{\sqrt{n}} = \frac{1}{\sqrt{9}} = 0.33$. We accept H_0 if $\bar{x} \leq 0$. Thus the probability of a Type II error when $\mu = 1$ is

$$P(\bar{x} \leq 0) = P\left(\frac{\bar{x} - 1}{0.33} \leq \frac{0 - 1}{0.33} \right) = P(Z \leq -3.0) = .0013$$

Exercise 16.48

We are using a significance level of 0.05. The probability of a Type I error is the same as the significance level: 0.05.

The probability of a Type II error when a specific alternative is true is the probability of accepting the null hypothesis. One minus this probability (the probability of accepting the null hypothesis) is the probability of (correctly) rejecting the null hypothesis. This last probability is the power at the specific alternative. We are given that this power is 0.78. One minus this value is thus the Type II error. Hence the Type II error is $1 - 0.78 = 0.22$.

CHAPTER 17

FROM EXPLORATION TO INFERENCE: PART II REVIEW

To assist you in reviewing the material in Chapters 8 – 16, we provide the text chapter and related problems in this Study Guide for each of the odd numbered review exercises. Other than pointing you in the right direction, we provide no additional hints or solutions. At this point, you should be able to work these problems on your own with minimal assistance. As a final challenge, we encourage you to work some of the Supplementary Exercises, which integrate more fully the material in these chapters.

Exercise 17.1
Text Location – Chapters 8 and 9 for observational studies and experiments, Chapters 8 and 9 for explanatory and response variables
Related Study Guide exercises – Exercises 8.1, 9.3, 9.8, 9.9

Exercise 17.3
Text Location – Chapter 8 for response variables, selecting an SRS
Related Study Guide exercises – Exercises 8.1, 8.10

Exercise 17.5
Text Location – Chapter 9 for designing an experiment, randomization, and response variable
Related Study Guide exercises – Exercises 9.3, 9.32

Exercise 17.7
Text Location – Chapter 11 for parameters and statistics
Related Study Guide exercise – Exercise 11.3

Exercise 17.9
Text Location – Chapter 8 for identifying bias in samples
Related Study Guide exercises – Exercises 8.31, 8.49

Exercise 17.11
Text Location – Chapter 15 for tests for a population mean
Related Study Guide exercise – Exercise 15.39

Exercise 17.13
Text Location – Chapter 15 for tests for a population mean, Chapter 16 for statistical significance and practical significance
Related Study Guide exercise – Exercises 15.39, 16.7

Exercise 17.15
Text Location – Chapter 15 for tests for a population mean
Related Study Guide exercise – Exercise 15.39

Exercise 17.17
Text Location – Chapter 14 for confidence intervals for the mean μ
Related Study Guide exercise – Exercise 14.25

Exercise 17.19
Text Location – Chapter 9 for observational studies, experiments, and confounding variables
Related Study Guide exercises – Exercises 9.8, 9.9

Exercise 17.21
(a) Text Location – Chapter 10 for probability rules
Related Study Guide exercise – Exercise 10.39

(b) Text Location – Chapter 10 for probability rules
Related Study Guide exercise – Exercise 10.39

Exercise 17.23
(a) Text Location – Chapter 10 for discrete probability models
Related Study Guide exercise – Exercise 10.39

(b) Text Location – Chapter 10 for discrete probability models
Related Study Guide exercise – Exercise 10.39

(c) Text Location – Chapter 10 for discrete probability models
Related Study Guide exercise – Exercise 10.39

(d) Text Location – Chapter 10 for random variables
Related Study Guide exercise – Exercise 10.39

Exercise 17.25
Text Location – Chapter 11 for the central limit theorem
Related Study Guide exercise – Exercise 11.13

Exercise 17.27
(a) Text Location – Chapter 3 for the 68–95–99.7 rule
Related Study Guide exercise – Exercise 3.7

(b) Text Location – Chapter 11 for the central limit theorem
Related Study Guide exercise – Exercise 11.13

Exercise 17.29
Text Location – Chapter 15 for tests for a population mean
Related Study Guide exercise – Exercise 15.39

Exercise 17.31
Text Location – Chapter 14 for confidence intervals for the mean μ
Related Study Guide exercise – Exercise 14.25

Exercise 17.33
Text Location – Chapter 9 for designing an experiment, randomization and response variable
Related Study Guide exercises – Exercises 9.3, 9.32

Exercise 17.35
a) Text Location – Chapter 15 for P-values
Related Study Guide exercises – Exercises 15.14, 15.52

b) Text Location – Chapter 15 for P-values
Related Study Guide exercises – Exercises 15.14, 15.52

Exercise 17.37
Text Location – Chapter 15 for P-values, Chapter 5 for facts about least-squares regression.
Related Study Guide exercises – Exercises 5.9, 15.14, 15.52

CHAPTER 18

INFERENCE ABOUT A POPULATION MEAN

OVERVIEW

Confidence intervals and significance tests for the mean μ of a normal population are based on the sample mean $\bar{x}$ of an SRS. When the sample size n is large, the central limit theorem suggests that these procedures are approximately correct for other population distributions. In Chapters 14 and 15 of your text, the (unrealistic) situation is considered in which we know the population standard deviation, σ. In this chapter, we consider the more realistic case where σ is not known and we must estimate σ from our SRS by the sample standard deviation s. In Chapters 15 and 16 we used the **one-sample z statistic**

$$z = \frac{\bar{x} - \mu}{\sigma/\sqrt{n}}$$

which has the $N(0,1)$ distribution. Replacing σ with s, we now use the **one-sample t statistic**

$$t = \frac{\bar{x} - \mu}{s/\sqrt{n}}$$

which has the **t distribution** with $n - 1$ **degrees of freedom**. For every positive value of k there is a t distribution with k degrees of freedom, denoted $t(k)$. All are symmetric, bell-shaped distributions, similar in shape to normal distributions but with greater spread. As k increases, $t(k)$ approaches the $N(0,1)$ distribution.

A level C **confidence interval for the mean** μ of a normal population when σ is unknown is

$$\bar{x} \pm t^* \frac{s}{\sqrt{n}}$$

where t^* is the upper $(1 - C)/2$ critical value of the $t(n - 1)$ distribution, whose value can be found in Table C in your text or from statistical software. The one-sample t confidence interval has the form estimate $\pm t^* SE_{estimate}$, where "SE" stands for **standard error**.

Significance tests of H_0: $\mu = \mu_0$ are based on the one-sample t statistic. P-values or fixed significance levels are computed from the $t(n - 1)$ distribution using Table C or, more commonly in practice, using statistical software.

One application of these one-sample t procedures is to the analysis of data from **matched pairs** studies. We compute the differences between the two values of a matched pair (often before and after measurements on the same unit) to produce a single sample value. The sample mean and standard deviation of these differences are computed. Depending on whether we are interested in a confidence interval or a test of significance concerning the difference in the population means of matched pairs, we use either the one-sample confidence interval or the one-sample significance test based on the t statistic.

For larger sample sizes, the t procedures are fairly **robust** against nonnormal populations. As a rule of thumb, t procedures are useful for nonnormal data when $n \geq 15$ unless the data show outliers or strong skewness, and for samples of size $n \geq 40$, t procedures can be used for even clearly skewed distributions. For smaller samples, it is a good idea to examine stemplots or histograms before you use the t procedures to check for outliers or skewness.

GUIDED SOLUTIONS

Exercise 18.7

KEY CONCEPTS: One-sample t confidence intervals, checking assumptions

The four-step process follow.

> *State.* What is the practical question in the context of the real-world setting?

> *Formulate.* What specific statistical operations does this problem call for?

> *Solve.* Make the graphs and carry out any calculations needed for this problem.

> *Conclude.* Give your practical conclusion in the setting of the real-world problem.

To apply the steps to this problem, here are some suggestions.

State. What characteristic of ancient air is of interest here? What question about this characteristic do we wish to answer?

Formulate. What method of inference will we use? What is the level of confidence?

Solve. Are the conditions for inference satisfied? (Do we have an SRS? Is the population approximately normal?) With a sample size of only $n = 9$, the most sensible graph for determining whether the population is approximately normal is probably a stemplot. Complete the stemplot that follows. Use split stems and use just the numbers to the left of the decimal place.

```
4 |
5 |
5 |
6 |
6 |
```

What do you conclude?

To compute a level C confidence interval, we use the formula $\bar{x} \pm t^* \dfrac{s}{\sqrt{n}}$, where t^* is the upper $(1 - C)/2$ critical value of the $t(n - 1)$ distribution, which can be found in Table C. Fill in the missing values. Don't forget to subtract 1 from the sample size when finding the appropriate degrees of freedom for the t confidence interval.

$C =$
$n =$
$t^* =$

Now compute the values of $\bar{x}$ and s from the data given. Use statistical software or a calculator.

$\bar{x} =$ $\qquad\qquad\qquad\qquad$ $s =$

Substitute all these values into the formula to complete the computation of the 95% confidence interval.

$\bar{x} \pm t^* \dfrac{s}{\sqrt{n}} =$

Conclude. State clearly what you have found in terms of the mean percent of nitrogen in ancient air.

Exercise 18.36

KEY CONCEPTS: Confidence intervals based on the one-sample t statistic, assumptions underlying t procedures

(a) To compute a level C confidence interval, we use the formula $\bar{x} \pm t^* \dfrac{s}{\sqrt{n}}$, where t^* is the upper $(1 - C)/2$ critical value of the $t(n - 1)$ distribution, which can be found in Table C. Fill in the missing values. Don't foget to subtract 1 from the sample size when finding the appropriate degrees of freedom for the t confidence interval.

$C =$
$n =$
$t^* =$

The values of $\bar{x}$ and s are given in the problem.

$\bar{x} =$ $\qquad\qquad\qquad\qquad$ $s =$

Substitute all these values into the formula to complete the computation of the 95% confidence interval.

$$\bar{x} \pm t^* \frac{s}{\sqrt{n}} =$$

(b) What are the assumptions required for the t confidence interval? Which assumptions are satisfied and which may not be? How were the subjects in the study obtained? How were the subjects in the placebo group obtained?

Exercise 18.41

KEY CONCEPTS: One-sample t tests

The four-step process follows.

   State. What is the practical question in the context of the real-world setting?

   Formulate. What specific statistical operations does this problem call for?

   Solve. Make the graphs and carry out any calculations needed for this problem.

   Conclude. Give your practical conclusion in the setting of the real-world problem.

To apply the steps to this problem, here are some suggestions.

State. What characteristic of the crankshafts is of interest here? What question about this characteristic do we wish to answer?

Formulate. What method of inference will we use? What is the parameter of interest?

State the appropriate null and alternative hypotheses in terms of this parameter.

H_0:

H_a:

Solve. Are the conditions for inference satisfied? (Do we have an SRS? Is the population approximately normal?)

To explore the normal distribution assumption, use the axes that follow to make a histogram of these data. Use class intervals 223.785–223.925, 223.925–223.975, and so on.

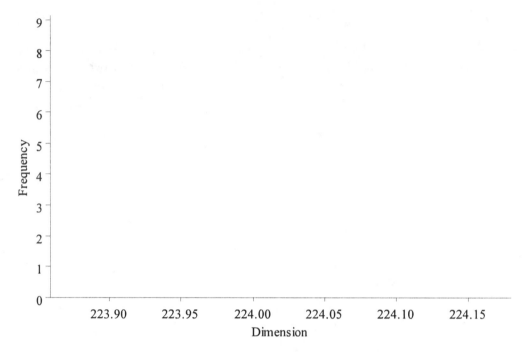

What do you conclude?

From the data, first calculate the sample mean and the sample standard deviation, preferably using statistical software or a calculator. Fill in the values.

$\bar{x} =$ $s =$

Use these values to compute the value of the one-sample t statistic.

$$t = \frac{\bar{x} - \mu_0}{s/\sqrt{n}} =$$

Compute the P-value using Table C or software. How many degrees of freedom are there? The degrees of freedom tell you which row of Table C you need to refer to for critical values. Remember, if the alternative is two-sided, then the probability found in the table needs to be doubled.

Degrees of freedom:

P-value:

Conclude. State clearly what you have found in terms of crankshaft dimension.

Exercise 18.43

KEY CONCEPTS: Matched pairs experiments, one-sample *t* tests

(a) This is a matched pairs experiment. The matched pair of observations are the right-hand and left-hand times on each subject. To avoid confounding with time of day, we would probably want subjects to use both knobs in the same session. We would also want to randomize which knob the subject uses first. How might you do this randomization? What about the order in which the subjects are tested?

(b) The four-step process follows.

State. What is the practical question in the context of the real-world setting?

Formulate. What specific statistical operations does this problem call for?

Solve. Make the graphs and carry out any calculations needed for this problem.

Conclude. Give your practical conclusion in the setting of the real-world problem.

To apply the steps to this problem, here are some suggestions.

State. What characteristic of the experiment is of interest here? What question about this characteristic do we wish to answer?

Formulate. The project hopes to show that right-handed people find right-hand threads easier to use than left-hand threads. In terms of the mean μ for the population of differences

$$\text{(left thread time)} - \text{(right thread time)}$$

what do we wish to show? This hypothesis would be the alternative. What are H_0 and H_a (in terms of μ)?

H_0:

H_a:

Solve. Are the conditions for inference satisfied? (Was the experiment properly randomized? Is the condition of normality satisfied?) For data from a matched pairs study, we compute the differences between the two values of a matched pair to produce a single sample value. These differences are as follows.

Right thread	Left thread	Difference = Left – Right
113	137	24
105	105	0
130	133	3
101	108	7
138	115	–23
118	170	52
87	103	16
116	145	29
75	78	3
96	107	11
122	84	–38
103	148	45
116	147	31
107	87	–20
118	166	48
103	146	43
111	123	12
104	135	31
111	112	1
89	93	4
78	76	–2
100	116	16
89	78	–11
85	101	16
88	123	35

Use the axes below to make a histogram of the differences. Use as class intervals –40 through –20, –20 through 0, and so on.

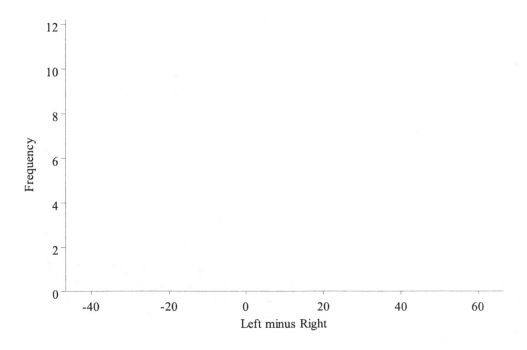

Does the normality condition appear to be satisfied?

The sample mean and the standard deviation of these differences need to be computed. Fill in their values. Use statistical software or a calculator.

$\bar{x} =$ $s =$

Now use the one-sample significance test based on the t statistic. What value of μ_0 should be used?

$$t = \frac{\bar{x} - \mu_0}{s/\sqrt{n}} =$$

From the value of the t statistic and Table C (or using statistical software), the P-value can be computed. Using Table C, between what two values does the P-value lie?

$\leq P\text{-value} \leq$

Exact P-value from software =

Note: This problem is most easily done directly using statistical software. The software will compute the differences, the t statistic and the P-value. Consult your user manual to see how to do one-sample t tests.

Conclude. State clearly what you have found in terms of mean time to move the indicator a fixed distance. Relate this to the original goal of the project, namely to show that right-handed people find right-hand threads easier to use than left-hand threads.

Exercise 18.45

KEY CONCEPTS: Matched pairs experiments, confidence intervals

Taking the 25 differences (left–right), we get the mean and standard deviation of the differences as $\bar{x} = 13.32$, $s = 22.94$ (see Exercise 18.43 in this Study Guide). To compute a level C confidence interval, use the formula

$$\bar{x} \pm t^* \frac{s}{\sqrt{n}} =$$

where t^* is the upper $(1 - C)/2$ critical value of the $t(n - 1)$ distribution, which can be found in Table C. Substitute all these values into the formula to complete the computation of the 95% confidence interval. Don't forget to subtract one from the sample size when finding the appropriate degrees of freedom for the t confidence interval.

As an alternative to computing the mean of the differences, you could evaluate the ratio of the mean time for right-hand threads as a percent of left-hand threads to help determine whether the time saved is of practical importance.

$\bar{x}_R / \bar{x}_L =$

COMPLETE SOLUTIONS

Exercise 18.7

State. We are interested in the mean percent of nitrogen in ancient air and we wish to estimate this quantity.

Formulate. We will estimate the mean percent of nitrogen in ancient air by giving a 90% confidence interval.

Solve. It is not clear that these data are an SRS from the late Cretaceous atmosphere, but we are told to assume that they are. The stemplot follows. There are no outliers, and the plot is slightly skewed left. With these few observations, it is difficult to check the assumptions. We might still use the t procedures but perhaps with not as much confidence in their validity as we had in other examples.

```
4 | 9
5 | 1 4
5 |
6 | 0 3 3 4 4
6 | 5
```

An approximate 90% confidence interval for the mean percent of nitrogen in ancient air can be calculated from the data on the 9 specimens of amber. We use the formula for a t interval, namely $\bar{x} \pm t^* \dfrac{s}{\sqrt{n}}$. In this problem, $C = 0.90$, $\bar{x} = 59.589$, $s = 6.2553$, $n = 9$; hence t^* is the upper $(1 - 0.90)/2 = 0.05$ critical value for the $t(8)$ distribution. From Table C we see that $t^* = 1.86$. Thus the 90% confidence interval is

$$59.589 \pm 1.86 \frac{6.2553}{\sqrt{9}} = 59.589 \pm 3.878 = (55.711, 63.467)$$

Many statistical software packages compute a confidence interval directly, after the data are entered.

Conclude. We are 90% confident that the mean percent of nitrogen in ancient air is between 55.711% and 63.467%.

Exercise 18.36

(a) A 95% confidence interval for the mean systolic blood pressure in the population from which the subjects were recruited can be calculated from the data on the 27 members of the placebo group, since they are randomly selected from the 54 subjects. We use the formula for a t interval, namely $\bar{x} \pm t^* \dfrac{s}{\sqrt{n}}$. In this exercise, $\bar{x} = 114.9$, $s = 9.3$, $n = 27$; hence t^* is the upper $(1 - 0.95)/2 = 0.025$ critical value for the $t(26)$ distribution. From Table C we see $t^* = 2.056$. Thus the 95% confidence interval is

$$114.9 \pm 2.056 \frac{9.3}{\sqrt{27}} = 114.9 \pm 3.68 = (111.22,\ 118.58)$$

(b) For the procedure used in (a), the population from which the subjects were drawn should be such that the distribution of the seated systolic blood pressure in the population is normal. The 27 subjects used for the confidence interval in part (a) should be a random sample from this population. Unfortunately, we do not know if that is the case. Although 27 subjects were selected at random from the total of 54 subjects in the study, we do not know if the 54 subjects were a random sample from this population.

With a sample of 27 subjects, it is not crucial that the population be normal, as long as the distribution is not strongly skewed and the data contain no outliers. It is important that the 27 subjects can be considered a random sample from the population. If not, we cannot appeal to the central limit theorem to ensure that the t procedure is at least approximately correct even if the data are not normal.

(Note: It turns out that since the subjects were divided at random into treatment and control groups, there do exist procedures for comparing the treatment and placebo groups. These procedures are not based on the t distribution, but they are valid as long as treatment groups are determined by randomization. However, the conclusions drawn from these procedures apply only to the subjects in the study. To generalize the conclusions to a larger population, we must know that the subjects are a random sample from this larger population.)

Exercise 18.41

State. We are interested in whether there is evidence that the mean crankshaft dimension is not 224 mm.

Formulate. We let μ represent the mean crankshaft dimension in the population of all crankshafts produced by the manufacturing process. We use a test of hypotheses to determine whether there is evidence that the mean crankshaft dimension is not 224 mm. We wish to test the hypotheses

$$H_0\text{: } \mu = 224 \text{ mm}$$
$$H_a\text{: } \mu \neq 224 \text{ mm}$$

Solve. We are not told whether the 16 crankshafts are an SRS from the population of those manufactured by the process. Thus, we have to assume that this is the case. In the histogram that follows, we can see that there are no outliers in the data. The data appear a bit skewed to the right but not strongly enough to threaten the validity of the t procedure given that the sample size is 16 (in the section on the robustness of t procedures, t procedures are safe for samples of size $n \geq 15$ unless there are outliers and/or strong skewness).

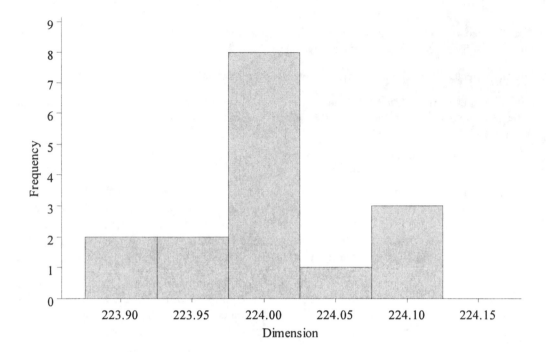

To test the hypotheses

$$H_0: \mu = 224 \text{ mm}$$
$$H_a: \mu \neq 224 \text{ mm}$$

we calculate the basic statistics $\bar{x} = 224.0019$, $s = 0.618$ and the standard error as $\dfrac{s}{\sqrt{n}} = \dfrac{0.0618}{\sqrt{16}} = 0.01545$. Substituting these in the formula for t yields

$$t = \frac{\bar{x} - \mu_0}{s/\sqrt{n}} = \frac{224.0019 - 224}{0.01545} = 0.123$$

The P-value for $t = 0.123$ is twice the area to the right of 0.123 under the t distribution curve with $n - 1 = 15$ degrees of freedom. Using Table C, we search the df = 15 row for entries that bracket 0.123. Since 0.123 lies to the left of the smallest entry in the table corresponding to a probability of 0.25, the P-value is

df = 15	
p	.25
t^*	0.691

therefore greater than .25 × 2 = 0.50 for this two-sided test. (Computer software gives a P-value of 0.9038).

Conclude. The data do not provide strong evidence that the mean crankshaft dimension differs from 224 mm.

Exercise 18.43

(a) The randomization might be carried out by simply flipping a fair coin. If the coin comes up heads, use the right-hand-threaded knob first. If the coin comes up tails, use the left-hand-threaded knob first. Alternatively, to balance the number of times each type is used first, one might choose an SRS of 12 of the 25 subjects. These 12 use the right-hand-threaded knob first. Everyone else uses the left-hand-threaded knob first.

A second place one might use randomization is in the order in which subjects are tested. Use a table of random digits to determine this order. Label subjects 01 to 25. The first label that appears in the list of random digits (read in groups of two digits) is the first subject measured; the second label that appears is the next subject measured; and so on. This randomization is probably less important than the one described in the previous paragraph. It would be important if the order or time at which a subject was tested might have an effect on the measured response. For example, if the study began early in the morning, the first subject might be sluggish if still sleepy. Sluggishness might lead to longer times and perhaps a larger difference in times. Subjects tested later in the day might be more alert.

(b) *State*. We are interested in whether right-handed people find right-hand threads easier to use than left-hand threads. The experiment actually measures the times in seconds each of 25 right-handed subjects took to move the indicator a fixed distance, once with the left-handed thread and once with the right-handed thread. Presumably shorter times indicate ease of use. Thus, we are interested in whether the times for the left-handed threads are greater than those for the right-handed threads.

Formulate. In terms of μ, the mean of the population of differences (left thread time) – (right thread time), we wish to test whether the times for the left-threaded knobs are longer than for the right-threaded knobs;

$$H_0: \mu = 0 \text{ and } H_a: \mu > 0$$

Solve. Assuming the randomization we recommended in part (a) is used, this would be a randomized experiment. A histogram of the 25 differences follows. We can see that there are no outliers in the data. The data appear a bit skewed to the left but not strongly enough to threaten the validity of the t procedure given that the sample size is 25 (in the section on the robustness of t procedures, t procedures are safe for samples of size $n \geq 15$ unless there are outliers and/or strong skewness).

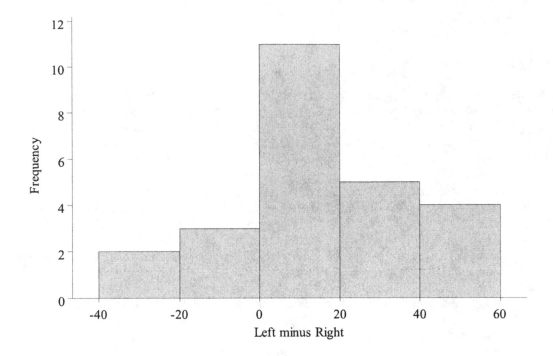

For the 25 differences we compute

$$\bar{x} = 13.32 \qquad\qquad s = 22.94$$

We then use the one sample significance test based on the t statistic.

$$t = \frac{\bar{x} - \mu_0}{s/\sqrt{n}} = \frac{13.32 - 0}{22.94/\sqrt{25}} = 2.903$$

From the value of the t statistic and Table C, the P-value is between 0.0025 and 0.005.

df = 24

p	.005	.0025
t^*	2.797	3.091

Using statistical software, the P-value is computed as P-value = 0.0039.

Conclude. We conclude that there is strong evidence that, on average, the time for left-hand threads is greater than the time for right-hand threads. Assuming that shorter times mean greater ease of use, we would conclude that there is strong evidence that the right-hand threads are easier to use.

Exercise 18.45

$\bar{x} = 13.32$, $s = 22.94$, $n = 25$, and t^* is the upper $(1 - 0.90)/2 = 0.05$ critical value for the $t(24)$ distribution. From Table D, we see that $t^* = 1.711$. Thus, the 90% confidence interval is

$$13.32 \pm 1.711 \frac{22.94}{\sqrt{25}} = 13.32 \pm 7.85 = (5.47, 21.17)$$

Computing the means, $\bar{x}_R = 104.12$, $\bar{x}_L = 117.44$, and $\bar{x}_R / \bar{x}_L = 88.7\%$, so people using the right-handed threads complete the task in about 90% of the time it takes those using the left-handed threads. As an alternative, if for each subject we first take the ratio right-thread/left-thread and then average these ratios, we get 91.7%, which is almost the same answer.

CHAPTER 19

TWO-SAMPLE PROBLEMS

OVERVIEW

One of the the most commonly used significance tests is the **comparison of two population means,** μ_1 and μ_2. In this setting we have two distinct, independent SRS's from two populations or two treatments in a randomized comparative experiment. The procedures are based on the difference $\bar{x}_1 - \bar{x}_2$. When the populations are not normal, the results obtained using the methods of this chapter are approximately correct due to the central limit theorem.

Tests and confidence intervals for the difference in the population means, $\mu_1 - \mu_2$, are based on the **two-sample t statistic.** Despite the name, this test statistic does *not* have an exact t distribution. However there are good approximations to its distribution that allow us to carry out valid significance tests. Conservative procedures use the $t(k)$ distribution as an approximation where the degrees of freedom k is taken to be the smaller of $n_1 - 1$ and $n_2 - 1$. More accurate procedures use the data to estimate the degrees of freedom k. This procedure is followed by most statistical software.

To carry out a significance test for H_0: $\mu_1 = \mu_2$, use the two-sample t statistic:

$$t = \frac{(\bar{x}_1 - \bar{x}_2)}{\sqrt{\dfrac{s_1^2}{n_1} + \dfrac{s_2^2}{n_2}}}$$

The P-value is found by using the approximate distribution $t(k)$, where k is estimated from the data when using statistical software or can be taken to be the smaller of n_1-1 and n_2-1 for a conservative procedure. An approximate confidence C level **confidence interval** for $\mu_1 - \mu_2$ is given by

$$(\bar{x}_1 - \bar{x}_2) \pm t^* \sqrt{\dfrac{s_1^2}{n_1} + \dfrac{s_2^2}{n_2}}$$

where t^* is the upper $(1 - C)/2$ critical value for $t(k)$, where k is estimated from the data when using statistical software or can be taken to be the smaller of $n_1 - 1$ and $n_2 - 1$ for a conservative procedure. The procedures are most robust to failures in the assumptions when the sample sizes are equal.

The **pooled two-sample t procedures** are used when we can safely assume that the two populations have equal variances. The modifications in the procedure are the use of the pooled estimator of the common unknown variance and critical values obtained from the $t\,(n_1 + n_2 - 2)$ distribution.

There are formal inference procedures for comparing the standard deviations of two normal populations as well as the two means. The validity of the procedures is seriously affected when the distributions are nonnormal, and they are not recommended for regular use. The procedures are based on the **F statistic,** which is the ratio of the two sample variances:

$$F = \frac{s_1^2}{s_2^2}$$

If the data consist of independent simple random samples of sizes n_1 and n_2 from two normal populations, then the F statistic has the F distribution, $F(n_1 - 1, n_2 - 1)$, if the two population standard deviations σ_1 and σ_2 are equal. Critical values of the F distribution are provided in Table D. Because of the skewness of the F distribution, when carrying out the two-sided test we take the ratio of the larger to the smaller standard deviation, which eliminates the need for lower critical values.

GUIDED SOLUTIONS

Exercise 19.3

KEY CONCEPTS: Single sample, matched pairs, or two samples

Are there one or two samples involved? Was matching done?

Exercise 19.4

KEY CONCEPTS: Single sample, matched pairs, or two samples

Are there one or two samples involved? Was matching done?

Exercise 19.5

KEY CONCEPTS: Single sample, matched pairs, or two samples

(a) Recall that SEM $= \dfrac{s}{\sqrt{n}}$ is the standard error of the mean. The description of the study provides the sample sizes. The means are given, and you should be able to reproduce the standard deviation s from the value of SEM and the sample size for each group. Write your answers in the table.

Group	Location	n	$\bar{x}$	s
1	Oregon			
2	California			

(b) Based on the two sample sizes, what value would you use for the degrees of freedom in the conservative two-sample t procedure?

Exercise 19.13

KEY CONCEPTS: Tests using the two-sample t, t approximation

The summary statistics obtained from the CrunchIt! output for the comparison of the measured variable (height of the second spike as a percent of the first) between poisoned and unpoisoned rats are presented in the following table.

Group	Treatment	n	$\bar{x}$	s
1	Poisoned	6	17.6000	6.3401
2	Unpoisoned	6	9.4998	1.9501

The values of the t statistic and the approximate degrees of freedom are provided in the output of CrunchIt! These summary statistics are all you need to verify the calculations.

$$t = \frac{(\bar{x}_1 - \bar{x}_2)}{\sqrt{\dfrac{s_1^2}{n_1} + \dfrac{s_2^2}{n_2}}}$$

$$df = \frac{\left(\dfrac{s_1^2}{n_1} + \dfrac{s_2^2}{n_2}\right)^2}{\dfrac{1}{n_1 - 1}\left(\dfrac{s_1^2}{n_1}\right)^2 + \dfrac{1}{n_2 - 1}\left(\dfrac{s_2^2}{n_2}\right)^2} =$$

Exercise 19.15

KEY CONCEPTS: Tests using the two-sample t, interpreting results

All of the calculations have been performed by CrunchIt! and are provided in the output. Using this information, write a summary in a sentence or two. Be sure to include the t, df, P, and a conclusion.

Exercise 19.21

KEY CONCEPTS: The F test for equality of the standard deviations of two normal populations

The summary statistics obtained from the CrunchIt! output for the comparison of the measured variable (height of the second spike as a percent of the first) between poisoned and unpoisoned rats are presented in the following table.

Group	Treatment	n	$\bar{x}$	s
1	Poisoned	6	17.6000	6.3401
2	Unpoisoned	6	9.4998	1.9501

We are interested in testing the hypotheses H_0: $\sigma_1 = \sigma_2$ and H_a: $\sigma_1 \neq \sigma_2$. The two-sided test statistic is the larger variance divided by the smaller variance, and if σ_1 and σ_2 are equal has the $F(n_1 - 1, n_2 - 1)$. Remember that n_1 is the numerator sample size. To organize your calculations, first compute the three quantities below.

$$F = \frac{\text{larger } s^2}{\text{smaller } s^2} =$$

Numerator df $(n_1 - 1) =$

Denominator df $(n_2 - 1) =$

Compare the value of the F you computed to the critical values given in Table D, making sure to go to the row and column corresponding to the appropriate degrees of freedom, or as close as you can get to these degrees of freedom if they are not in the table. Between which two critical values does F lie? What can be said about the P-value from the table? If available, use software to calculate the exact P-value.

Exercise 19.34

KEY CONCEPTS: Tests using the two-sample t, checking assumptions, back-to-back stemplots

The four-step process follows.

State. What is the practical question, in the context of the real-world setting?

Formulate. What specific statistical operations does this problem call for?

Solve. Make the graphs and carry out any calculations needed for this problem.

Conclude. Give your practical conclusion in the setting of the real-world problem.

To apply these to this problem, here are some suggestions.

State. What characteristic of SSHA scores is of interest here? What question about this characteristic for men and women do we wish to answer?

Formulate. Letting μ_W and μ_M represent the population means for women and men respectively, first state the hypotheses of interest in terms of these parameters.

H_0:

H_a:

Solve. Are the conditions for inference satisfied? (Do we have an SRS? Is the population approximately normal?) For the relatively small samples in this problem, we might use a back-to-back stemplot to examine the data. Complete the stemplot below. We have filled in the three smallest men's and the three smallest women's scores. Are there outliers and/or skewness? Are t procedures still appropriate? Explain.

```
    Women        Men
             7 | 05
             8 | 8
             9 |
       931 |10 |
            11 |
            12 |
            13 |
            14 |
            15 |
            16 |
            17 |
            18 |
            19 |
            20 |
```

Below are the summary statistics. You can use a calculator or statistical software to verify these values. Now use these summary statistics to compute the value of t below.

Group	n	$\bar{x}$	s
Women	18	141.056	26.4363
Men	20	121.250	32.8519

$$t = \frac{\bar{x}_W - \bar{x}_M}{\sqrt{\dfrac{s_W^2}{n_W} + \dfrac{s_M^2}{n_M}}} =$$

Using the conservative degrees of freedom, what can you say about the P-value?

Note that if you did this problem by entering the data into statistical software, the value of the t statistic should be the same, but the degrees of freedom used will be different and the P-value will differ slightly.

Conclude. What do you conclude about the SSHA scores of first-year men versus first-year women at this private college?

Exercise 19.39

KEY CONCEPTS: Two-sample t confidence interval

The formula for the 90% confidence interval is $(\bar{x}_1 - \bar{x}_2) \pm t^* \sqrt{\dfrac{s_1^2}{n_1} + \dfrac{s_2^2}{n_2}}$, where t^* is the upper $(1 - C)/2 =$ 0.05 critical value for the t distribution with degrees of freedom equal to the smaller of $n_1 - 1$ and $n_2 - 1$. The sample sizes, means, and SEMs are given in the problem. Recall that SEM $= \dfrac{s}{\sqrt{n}}$ so that $s =$ SEM $\sqrt{n}$. Don't forget to square the standard deviations in the formula for the standard error. Complete the calculations in steps as suggested.

$n_1 =$

$n_2 =$

$s_1 = \text{SEM}_1 \sqrt{n_1} =$

$s_2 = \text{SEM}_2 \sqrt{n_2} =$

$t^* =$

$\sqrt{\dfrac{s_1^2}{n_1} + \dfrac{s_2^2}{n_2}} =$

$(\bar{x}_1 - \bar{x}_2) \pm t^* \sqrt{\dfrac{s_1^2}{n_1} + \dfrac{s_2^2}{n_2}} =$

What is the relation between confidence intervals and tests of significance?

COMPLETE SOLUTIONS

Exercise 19.3

This example involves a single sample. We have a sample of 20 measurements, and we want to see if the mean for this method agrees with the known concentration.

Exercise 19.4

This example involves two samples, the set of measurements on each method. Note that we are not told of any matching.

Exercise 19.5

(a) Measurements were taken at 6 locations in Oregon and 7 locations in California. The sample sizes are then 6 for Oregon and 7 for California. The means are given, and since SEM $= \dfrac{s}{\sqrt{n}}$, the standard deviation for each group is $s = $ SEM $\sqrt{n}$. For Oregon this gives $s = $ SEM $\sqrt{n} = 1.56\sqrt{6} = 3.82$, and for California $s = $ SEM $\sqrt{n} = 2.68\sqrt{7} = 7.09$.

Group	Location	n	$\bar{x}$	s
1	Oregon	6	26.9	3.82
2	California	7	11.9	7.09

(b) The degrees of freedom are the smaller of $n_1 - 1 = 6 - 1 = 5$ and $n_2 - 1 = 7 - 1 = 6$, which is 5.

Exercise 19.13

Entering the summary statistics into the formulas for the t statistic and the approximate degrees of freedom gives

$$t = \frac{\bar{x}_1 - \bar{x}_2}{\sqrt{\dfrac{s_1^2}{n_1} + \dfrac{s_2^2}{n_2}}} = \frac{17.6000 - 9.4998}{\sqrt{\dfrac{(6.3401)^2}{6} + \dfrac{(1.9501)^2}{6}}} = 2.99$$

and

$$df = \frac{\left(\dfrac{s_1^2}{n_1} + \dfrac{s_2^2}{n_2}\right)^2}{\dfrac{1}{n_1 - 1}\left(\dfrac{s_1^2}{n_1}\right)^2 + \dfrac{1}{n_2 - 1}\left(\dfrac{s_2^2}{n_2}\right)^2} = \frac{\left(\dfrac{6.3401^2}{6} + \dfrac{1.9501^2}{6}\right)^2}{\dfrac{1}{6 - 1}\left(\dfrac{6.3401^2}{6}\right)^2 + \dfrac{1}{6 - 1}\left(\dfrac{1.9501^2}{6}\right)^2} = 5.9$$

Exercise 19.15

Using the information in the CrunchIt! output, we have $t = 2.9912$, df $= 5.9$ and the P-value $= 0.0246$. We would reject the null hypothesis at a significance level of 0.05 since the P-value falls below 0.05 but not at the 0.01 significance level. We conclude that there is good evidence of a difference in the mean of the measured variable (height of the second spike as a percent of the first) between poisoned and unpoisoned rats, with the poisoned rats tending to have larger values of the measured variable.

Exercise 19.21

Since the treated group has the larger standard deviation (and variance), this is the variance that goes into the numerator, so the degrees of freedom are $n_1 - 1 = 6 - 1 = 5$ and $n_2 - 1 = 6 - 1 = 5$. To compute the value of the test statistic, note that it is the standard deviations, not the variances, that are given as summary statistics. The standard deviations must first be squared to get the two variances. The larger standard deviation is for the poisoned group and the sample variance for this group is $(6.3401)^2 = 40.1969$. For the untreated group, the sample variance is $(1.9501)^2 = 3.8029$. The value of the test statistic is

$$F = \frac{\text{larger } s^2}{\text{smaller } s^2} = \frac{40.1969}{3.8029} = 10.57$$

Since the degrees of freedom (5, 5) are in the table, we go to Table D and find the two values that bracket the computed value of $F = 8.68$.

df = (5, 5)		
p	0.025	0.01
F^*	7.15	10.97

Since the test is two-sided, we double the significance levels from the table and conclude that the P-value is between 0.02 and 0.05. Using computer software, we find the P-value to be 0.0216893. This is significant evidence of unequal standard deviations between the two groups at the 5% level but not at the 1% level.

Exercise 19.34

State. We are interested in whether among first-year students at the private college, the mean SSHA score for men is lower than the mean score for a comparable group of women.

Formulate. We test H_0: $\mu_W = \mu_M$ and H_a: $\mu_W > \mu_M$.

Solve. We are told that the data are an SRS. Both stemplots show distributions that are slightly skewed to the right, and each has one or two moderate high outliers. Since the sum of the sample sizes is close to 40, a t procedure may be used but with some caution.

```
     Women        Men
               7 | 05
               8 | 8
               9 | 12
         931  10 | 489
           5  11 | 3455
         966  12 | 6
          77  13 | 2
          80  14 | 06
         442  15 | 1
          55  16 | 9
           8  17 |
              18 | 07
              19 |
           0  20 |
```

From the summary statistics, the value of t is computed as

$$t = \frac{\bar{x}_W - \bar{x}_M}{\sqrt{\dfrac{s_W^2}{n_W} + \dfrac{s_M^2}{n_M}}} = \frac{141.056 - 121.250}{\sqrt{\dfrac{(26.4363)^2}{18} + \dfrac{(32.8519)^2}{20}}} = 2.06$$

Since the degrees of freedom are the smaller of $n_W - 1 = 17$ and $n_M - 1 = 19$, we go to Table C using df = 17 and find the two values that bracket the computed value of $t = 2.06$.

	df = 17	
p	0.025	0.05
t^*	2.110	1.740

Because the test is one-sided, $0.025 < P\text{-value} < 0.05$. Statistical software gives a P-value of 0.02752375.

Conclude. The exercise yields moderately strong evidence (statistically significant at the 5% level) that first-year men have lower mean SSHA scores than do first-year women at this private college.

Exercise 19.39

We compute

$n_1 = 9$

$n_2 = 11$

$s_1 = \text{SEM}_1 \sqrt{n_1} = 7\sqrt{9} = 21$

$s_2 = \text{SEM}_2 \sqrt{n_2} = 10\sqrt{11} = 33.16625$

The upper $(1 - C)/2 = 0.05$ critical value for the t distribution with degrees of freedom equal to the smaller of $n_1 - 1 = 8$ and $n_2 - 1 = 10$ is

$t^* = 1.86$

$$\sqrt{\frac{s_1^2}{n_1} + \frac{s_2^2}{n_2}} = \sqrt{\frac{21^2}{9} + \frac{33.16625^2}{11}} = \sqrt{149} = 12.2$$

$$(\bar{x}_1 - \bar{x}_2) \pm t^* \sqrt{\frac{s_1^2}{n_1} + \frac{s_2^2}{n_2}} = (59 - 32) \pm 1.86 \times 12.2 = 27 \pm 22.692 = (4.308, 49.692).$$

Recall that a level α two-sided significance test rejects a hypothesis H_0: $\mu_1 - \mu_2 = 0$ at level α exactly when 0 falls outside a level $1 - \alpha$ confidence interval for $\mu_1 - \mu_2$. This $(1 - \alpha) \times 100\% = 90\%$ confidence interval does not include 0; hence we would not reject H_0: $\mu_1 - \mu_2 = 0$ in favor of H_a: $\mu_1 - \mu_2 \neq 0$ at level α $= (1 - 0.90) = 0.10$.

CHAPTER 20

INFERENCE ABOUT A POPULATION PROPORTION

OVERVIEW

In this chapter we consider inference about a population proportion p based on the **sample proportion**

$$\hat{p} = \frac{\text{count of successes in the sample}}{\text{count of observations in the sample}}$$

obtained from an SRS of size n, where X is the number of "successes" (occurrences of the event of interest) in the sample. To use the methods of this chapter for inference, the following assumptions need to be satisfied.

• The data are an SRS from the population of interest.

• The population is at least 10 times as large as the sample.

• The sample size is sufficiently large. Guidelines for sample sizes are given.

In this case, we can treat $\hat{p}$ as having a distribution that is approximately normal with mean $\mu = p$ and standard deviation $\sigma = \sqrt{p(1-p)/n}$.

An **approximate level C confidence interval** for p is

$$\hat{p} \pm z^* \sqrt{\frac{\hat{p}(1-\hat{p})}{n}}$$

where z^* is the upper $(1-C)/2$ critical value of the standard normal distribution,

$$\sqrt{\frac{\hat{p}(1-\hat{p})}{n}}$$

is the **standard error** of $\hat{p}$, and $z^* \sqrt{\dfrac{\hat{p}(1-\hat{p})}{n}}$ is the **margin of error.** Use this interval only when the counts of successes and failures in the sample are both at least 15.

A more accurate confidence interval for smaller samples is the **plus four confidence interval**. To get this interval, add four imaginary observations, two successes and two failures, to your sample. Then, with these new values for the number of failures and successes, use the previous formula for the approximate level C confidence interval. Use the plus four confidence interval when the confidence level C is at least 90% and the sample size n is at least 10.

The **sample size** n required to obtain a confidence interval of approximate margin of error m for a proportion is

$$n = \left(\frac{z^*}{m}\right)^2 p^*(1-p^*)$$

where p^* is a guessed value for the population proportion and z^* is the upper $(1-C)/2$ critical value of the standard normal distribution. To guarantee that the margin of error of the confidence interval is less than or equal to m no matter what the value of the population proportion may be, use a guessed value of $p^* = 1/2$, which yields

$$n = \left(\frac{z^*}{2m}\right)^2.$$

Tests of the hypothesis $H_0: p = p_0$ are based on the z **statistic**.

$$z = \frac{\hat{p} - p_0}{\sqrt{\dfrac{p_0(1-p_0)}{n}}}$$

with P-values calculated from the $N(0, 1)$ distribution. Use this test when $np_0 \geq 10$ and $n(1-p_0) \geq 10$.

GUIDED SOLUTIONS

Exercise 20.1

KEY CONCEPTS: Parameters and statistics, proportions

(a) To what group does the study refer?

Population =

Parameter p =

(b) A statistic is a number computed from a sample. What is the size of the sample and how many in the sample said they prayed at least once in a while? From these numbers compute

$\hat{p}$ =

Exercise 20.6

KEY CONCEPTS: When to use the procedures for inference about a proportion

Recall the assumptions needed to safely use the methods of this chapter to compute a confidence interval:

- The data are an SRS from the population of interest.

- The population is at least 10 times as large as the sample.

- For a confidence interval, n is large enough that both the count of successes $n\hat{p}$ and the count of failures $n(1 - \hat{p})$ are 15 or more.

These are the conditions we must check. Are all the conditions met?

Exercise 20.9

KEY CONCEPTS - large sample confidence intervals for a proportion

(a) We know that the proportion of successes is given by the formula

$$\hat{p} = \frac{\text{count of successes in the sample}}{\text{count of observations in the sample}}$$

We are interested in finding the number among the 1009 interviewed who thought deaths from guns would increase. This is the count of successes in the sample. We know that

$$\hat{p} =$$

and

$$n =$$

Use these numbers and the formula to find the number among the 1009 interviewed who thought deaths from guns would increase.

(b) Explain to someone who knows no statistics what "margin of error plus or minus 3 percentage points" means.

(c) Recall the assumptions needed to safely use the methods of this chapter to compute a confidence interval:

- The data are an SRS from the population of interest.

- The population is at least 10 times as large as the sample.
- For a confidence interval, n is large enough that both the count of successes and the count of failures are 15 or more.

These are the conditions we must check. Are all the conditions met?

An approximate 95% confidence interval for the proportion p of all adults who think that deaths from guns will increase is given by the formula

$$\hat{p} \pm z^* \sqrt{\frac{\hat{p}(1-\hat{p})}{n}}$$

where z^* is the upper 0.025 critical value of the standard normal distribution. What are n, $\hat{p}$, and z^* here?

n = sample size =

$\hat{p}$ = the proportion of adults in the sample who think that deaths from guns will increase =

z^* =

Compute the 95% confidence interval.

$$\hat{p} \pm z^* \sqrt{\frac{\hat{p}(1-\hat{p})}{n}} =$$

Does your margin of error agree with the 3 percentage points announced by Harris?

Exercise 20.14

KEY CONCEPTS: Sample size and margin of error

The sample size n required to obtain a confidence interval of approximate margin of error m for a proportion is

$$n = \left(\frac{z^*}{m}\right)^2 p^*(1-p^*)$$

where p^* is a guessed value for the population proportion and z^* is the critical value of the standard normal distribution for the desired level of confidence. To apply this formula here we must determine

m = desired margin of error =

p^* = a guessed value for the population proportion =

C = desired level of confidence =

z^* = the upper (1 C)/2 critical value of the standard normal distribution =

From the statement of the exercise, what are these values? Once you have determined them, use the formula to compute the required sample size n.

$$n = \left(\frac{z^*}{m}\right)^2 p^*(1 \; p^*) =$$

Exercise 20.17

KEY CONCEPTS: When to use the normal approximation to the binomial test

Recall that the rule of thumb is that the normal approximation to the binomial is appropriate if *both* $np_0 \geq 10$ and $n(1 - p_0) \geq 10$. These are the conditions we must check in (a) and (b).

(a)

(b)

Exercise 20.29

KEY CONCEPTS: Accurate confidence intervals for a proportion

(a) We use the plus four confidence interval. To get this interval, first add four imaginary observations, two successes and two failures, to the sample. The number of observations in the sample is 127 and the number of successes is 107 (number that said they prayed at least a few times a year). Thus, we pretend that

$n = 127 + 4 =$

number of successes (number that said they prayed at least a few times a year) = $107 + 2 =$

Now we use the formula for the confidence interval. An approximate 99% confidence interval for $p =$ the proportion of all students who pray is

$$\hat{p} \pm z^* \sqrt{\frac{\hat{p}(1-\hat{p})}{n}}$$

where z^* is the upper 0.005 critical value of the standard normal distribution. What are n, $\hat{p}$, and z^* here?

n = the "imaginary" sample size from above =

$\hat{p}$ = ("imaginary" number who said they prayed at least a few times a year)/n =

z^* =

Now compute the 99% confidence interval.

$$\hat{p} \pm z^* \sqrt{\frac{\hat{p}(1-\hat{p})}{n}} =$$

(b) Is it reasonable to assume that psychology and communication majors are representative of all undergraduates?

Exercise 20.33

KEY CONCEPTS: Testing hypotheses about a proportion

The four step process involves the following steps.

State: What is the practical question in the context of the real-world setting?

Formulate: What specific statistical operations does this problem call for?

Solve: Make the graphs and carry out any calculations needed for this problem.

Conclude: Give your practical conclusion in the setting of the real-world problem.

To apply the steps to this problem, here are some suggestions. You can use Example 20.7 in the text as a guide.

State: State the problem.

Formulate: What is the parameter of interest? What statistical hypotheses should you test to answer this question?

Solve: First you should check that the appropriate conditions for inference are satisfied:

Is the sample an SRS from a large population?

Is the sample size reasonably large?

Are $np_0 \geq 10$ and $n(1 \ p_0) \geq 10$?

Compute the sample proportion of female Hispanic drivers in Boston who wear seatbelts.

$\hat{p} =$

Compute

$$z = \frac{\hat{p} - p_0}{\sqrt{\dfrac{p_0(1 - p_0)}{n}}} =$$

and

P-value $=$

Conclude: What do you conclude?

COMPLETE SOLUTIONS

Exercise 20.1

(a) The population is presumably all college students. The parameter p is the proportion of all college students who pray at least once in a while.

(b) The statistic is $\hat{p}$ the proportion in the sample who said that they prayed at least once in a while.

$$\hat{p} = 107/127 = 0.8425$$

Exercise 20.6

The data are an SRS from the population of interest.
The population consists of 175 students. This is *not* at least 10 times as large as the sample size $n = 50$. Thus, the methods of this chapter *cannot* be safely used.

Exercise 20.9

(a) We know that the proportion of successes is given by the formula

$$\hat{p} = \frac{\text{count of successes in the sample}}{\text{count of observations in the sample}}$$

where $\hat{p} = 0.70$ and $n =$ count of observations in the sample $= 1009$. The count of successes in the sample is then $n\hat{p} = (0.70)(1009) = 706.3$. Rounding off to the nearest integer, the number of adults in the sample who think that deaths from guns will increase is 706.

(b) For the sample size we have taken, we expect that the proportion of the sample who think that deaths from guns will increase should not differ from the proportion in the population by more than 3%.

(c) The data are a random sample from the population of interest, all adults in the U.S. The total number of adults in the U.S. is at least 10 times as large as the sample size of $n = 1009$.

The sample size is $n = 1009$, which is large.

The number of successes (those who thought deaths from guns would increase) is 706, and the number of failures (those who didn't think deaths from guns would increase) is $1009 - 706 = 303$. Both of these are much bigger than 15.

The appropriate conditions for a large sample confidence interval are satisfied.

An approximate 95% confidence interval for $p =$ the proportion of all adults who think that deaths from guns will increase is

$$\hat{p} \pm z^* \sqrt{\frac{\hat{p}(1 - \hat{p})}{n}}$$

where z^* is the upper 0.025 critical value of the standard normal distribution, which in this case is 1.96. We have $n = 1009$ and we know that the sample proportion of all adults who think that deaths from guns will increase is

$$\hat{p} = 0.70$$

and our 95% confidence interval is

$$\hat{p} \pm z^* \sqrt{\frac{\hat{p}(1 - \hat{p})}{n}} = 0.70 \pm 1.96 \sqrt{\frac{0.70(1 - 0.70)}{1009}} = 0.66 \pm 1.96(0.0144) = 0.70 \pm 0.03$$

or 0.67 to 0.73.

The margin of error quoted by Harris is 3%. Our margin of error from part is 0.03, which is 3% when converted to a percent. Thus, our results agree with the claim by the Harris poll.

Exercise 20.14

We start with the guess that $p* = 0.75$. For 95% confidence we use $z* = 1.96$. The sample size we need for a margin of error $m = 0.04$ is thus

$$n = \left(\frac{z*}{m}\right)^2 p*(1 - p*) = \left(\frac{1.96}{0.04}\right)^2 0.75(1 - 0.75) = 450.1875.$$

We round up to $n = 451$. Thus, a sample of size 451 is needed to estimate the proportion of Americans with at least one Italian grandparent who can taste PTC to within ± 0.04 with 95% confidence.

Exercise 20.17

(a) We see that $np_0 = (10)(0.5) = 5 < 10$, so the normal approximation to the binomial should *not* be used in this case.

(b) We see that $np_0 = (200)(0.99) = 198 \geq 10$ and $n(1 - p_0) = (200)(1 - 0.99) = (200)(0.01) = 2 < 10$. The normal approximation to the binomial should *not* be used in this case.

Exercise 20.29

(a) We use the plus four confidence interval. To get this interval, first add four imaginary observations, two successes and two failures, to the sample. This means that we pretend that

$n = 127 + 4 = 131$

number of successes (number that said they prayed at least a few times a year) $= 107 + 2 = 109$.

Now we use the formula for the confidence interval. An approximate 99% confidence interval for $p =$ the proportion of all students who pray is

$$\hat{p} \pm z* \sqrt{\frac{\hat{p}(1 - \hat{p})}{n}}$$

where $z*$ is the upper 0.005 critical value of the standard normal distribution, which in this case is $z* = 2.576$. We pretend $n = 131$ and that 109 of the respondents said they prayed at least a few times a year. Thus, the proportion of respondents who said they prayed at least a few times a year is

$$\hat{p} = 109/131 = 0.83$$

and our 99% confidence interval is

$$\hat{p} \pm z* \sqrt{\frac{\hat{p}(1 - \hat{p})}{n}} = 0.83 \pm 2.576 \sqrt{\frac{0.83(1 - 0.83)}{131}} = 0.83 \pm 2.576(0.0328) = 0.83 \pm 0.08$$

or 0.75 to 0.91.

(b) Students choose to be psychology or communication majors for a reason. Their interests and attitudes are likely to differ from those of students who choose other majors. It is probably unreasonable to assume that this is an SRS from the population of all students.

Exercise 20.33

State: We would like to know if more than 50% of Hispanic female drivers in Boston wear seatbelts. Investigators observed a random sample of 117 Hispanic female drivers and found that 68 of these drivers were wearing seatbelts. In our sample, the proportion of Hispanic female drivers wearing seatbelts was

$$\hat{p} = 68/117 = 0.5812,$$

or 58.12% of the sample. Although slightly more than half the sample were wearing seatbelts, does this sample provide evidence that more than 50% of all Hispanic female drivers wear seatbelts?

Formulate: Let p be the proportion of all Hispanic female drivers in Boston who wear seatbelts? We want to test the hypotheses

$$H_0: p = 0.5 \qquad H_a: p > 0.5$$

Solve: The sample is assumed to be a random sample of all Hispanic female drivers in Boston. The sample of $n = 117$ drivers is reasonably large and both $np_0 = (117)(0.5) = 58.5 \geq 10$ and $n(1 - p_0) = 117(1 - 0.5) = 58.5 \geq 10$, so the conditions for inference are met.

The sample proportion of female Hispanic drivers in Boston who wear seatbelts is $\hat{p} = 0.5812$ and

$$z = \frac{\hat{p} - p_0}{\sqrt{\dfrac{p_0(1 - p_0)}{n}}} = \frac{0.5812 - 0.5}{\sqrt{\dfrac{0.5(1 - 0.5)}{117}}} = \frac{0.0812}{0.0462} = 1.76.$$

The *P*-value is the area under the standard normal curve to the right of $z = 1.76$ which is $1 - 0.9608 = 0.0392$.

Conclude: There is moderate evidence that more than half of all Hispanic female drivers in Boston wear seatbelts.

CHAPTER 21

COMPARING TWO PROPORTIONS

OVERVIEW

Confidence intervals and tests designed to compare two population proportions are based on the **difference in the sample proportions** $\hat{p}_1 - \hat{p}_2$. The formula for the level C confidence interval is

$$\hat{p}_1 - \hat{p}_2 \pm z^*\text{SE}$$

where z^* is the upper $(1 - C)/2$ standard normal critical value and SE is the standard error for the difference in the two proportions computed as

$$\text{SE} = \sqrt{\frac{\hat{p}_1(1-\hat{p}_1)}{n_1} + \frac{\hat{p}_2(1-\hat{p}_2)}{n_2}}$$

In practice, use this confidence interval when the populations are at least 10 times as large as the samples and the counts of successes and failures are 10 or more in both samples.

To get a more accurate confidence interval for smaller samples, add four imaginary observations, one success and one failure in each sample. Then, with these new values for the number of failures and successes, use the previous formula for the approximate level C confidence interval. This is the **plus four confidence interval.** You can use it whenever both samples have five or more observations.

Significance tests for the equality of the two proportions, H_0: $p_1 = p_2$, use a different standard error for the difference in the sample proportions, which is based on a **pooled estimate** of the common (under H_0) value of p_1 and p_2,

$$\hat{p} = \frac{\text{count of successes in both samples combined}}{\text{count of observations in both samples combined}}$$

The test uses the z statistic

$$z = \frac{\hat{p}_1 - \hat{p}_2}{\sqrt{\hat{p}(1-\hat{p})\left(\dfrac{1}{n_1} + \dfrac{1}{n_2}\right)}}$$

and P-values are computed using Table A of the standard normal distribution. In practice, use this test when the populations are at least 10 times as large as the samples and the counts of successes and failures are five or more in both samples.

GUIDED SOLUTIONS

Exercise 21.19

KEY CONCEPTS: Testing equality of two population proportions

First verify that it is safe to use the z test for equality of two proportions.

Let p_1 represent the proportion of papers without statistical assistance that were rejected without being reviewed in detail and p_2 the proportion of papers with statistical help that were rejected without being reviewed in detail. Recall that a test of the hypothesis H_0: $p_1 = p_2$ uses the z statistic

$$z = \frac{\hat{p}_1 - \hat{p}_2}{\sqrt{\hat{p}(1-\hat{p})\left(\dfrac{1}{n_1} + \dfrac{1}{n_2}\right)}}$$

where n_1 and n_2 are the sizes of the samples, $\hat{p}_1$ and $\hat{p}_2$ are the estimates of p_1 and p_2, and

$$\hat{p} = \frac{\text{count of successes in both samples combined}}{\text{count of observations in both samples combined}}$$

First state the hypotheses to be tested. (What is the alternative in this case, one-sided or two-sided?)

The two sample sizes are

$n_1 =$

$n_2 =$

From the data, the estimates of these two proportions are

$\hat{p}_1 =$

$\hat{p}_2 =$

Compute

$$\hat{p} = \frac{\text{count of successes in both samples combined}}{\text{count of observations in both samples combined}} =$$

and then

$$z = \frac{\hat{p}_1 - \hat{p}_2}{\sqrt{\hat{p}(1-\hat{p})\left(\dfrac{1}{n_1} + \dfrac{1}{n_2}\right)}} =$$

Using Table A, compute

       P-value =

What do you conclude?

Exercise 21.21

KEY CONCEPTS: Large sample confidence intervals for the difference between two population proportions

First determine whether the conditions for the large sample confidence interval arc met or whether the plus four confidence interval needs to be used.

The two populations are proportions of papers rejected without review when a statistician is and is not involved in the research. The two sample sizes are

      n_1 = number of papers rejected without review without a statistician involved =

      n_2 = number of papers rejected without review with a statistician involved =

and the number of "successes" are

      Number of papers in sample rejected without review without a statistician involved =

      Number of papers in sample rejected without review with a statistician involved =

From the data, the estimates of the two proportions are

$\hat{p}_1 =$

$\hat{p}_2 =$

Let p_1 represent the proportion of all papers rejected without review without a statistician involved, and p_2 represent the proportion of all papers rejected without review with a statistician involved. Recall that a level C confidence interval for $p_1 - p_2$ is

$$(\hat{p}_1 - \hat{p}_2) \pm z^*\text{SE}$$

where z^* is the upper $(1 - C)/2$ standard normal critical value and SE is the standard error for the difference in the two proportions computed as

$$\text{SE} = \sqrt{\frac{\hat{p}_1(1-\hat{p}_1)}{n_1} + \frac{\hat{p}_2(1-\hat{p}_2)}{n_2}}$$

Use the values of $\hat{p}_1$ and $\hat{p}_2$ you computed to obtain the standard error:

$$\text{SE} = \sqrt{\frac{\hat{p}_1(1-\hat{p}_1)}{n_1} + \frac{\hat{p}_2(1-\hat{p}_2)}{n_2}} =$$

For a 95% confidence interval,

$$z^* =$$

Compute the interval:

$$(\hat{p}_1 - \hat{p}_2) \pm z^*\text{SE} =$$

Exercise 21.29

KEY CONCEPTS: Plus four confidence intervals for the difference between two population proportions

The four-step process follows.

   State: What is the practical question in the context of the real-world setting?

   Formulate: What specific statistical operations does this problem call for?

   Solve: Make the graphs and carry out any calculations needed for this problem.

   Conclude: Give your practical conclusion in the setting of the real-world problem.

To apply the steps to this problem, here are some suggestions. You may want to use Example 21.3 of the text as a guide.

State: In the space, describe the problem and the data obtained.

Formulate: Formulate the problem as either a confidence interval or hypothesis test.

Solve: We can use the large-sample confidence interval when the populations are at least 10 times as large as the samples and the counts of successes and failures are 10 or more in both samples.   Are these conditions satisfied?   If not, you must use the plus four confidence interval rather than the large-sample confidence interval.   Give the appropriate data summary below.

Whether you are using the plus four confidence interval or the large-sample interval, you must first compute the standard error:

SE =

For a 90% confidence interval, $z^* - 1.645$, so our 90% confidence interval is

$$(\hat{p}_1 - \hat{p}_2) \pm z^* \text{SE} =$$

Conclude: What can you conclude about the difference in proportions between the two areas?

Exercise 21.32

KEY CONCEPTS: Testing equality of two population proportions

The four step process follows.

  State: What is the practical question in the context of the real-world setting?

Formulate: What specific statistical operations does this problem call for?

Solve: Make the graphs and carry out any calculations needed for this problem.

Conclude: Give your practical conclusion in the setting of the real-world problem.

To apply the steps to this problem, here are some suggestions. You may want to use Example 21.4 of the text as a guide.

State: Describe the problem of interest and the data obtained.

Formulate: Formulate the problem as either a confidence interval or hypothesis test. If a hypothesis test is appropriate, is the alternative one- or two-sided?

Solve: First check the conditions for using the test.

To compute the test, first give the two sample sizes.

$n_1 =$

$n_2 =$

From the data, the estimates of the two proportions are

$\hat{p}_1 =$

$\hat{p}_2 =$

Compute

$$\hat{p} = \frac{\text{count of successes in both samples combined}}{\text{count of observations in both samples combined}} =$$

and then

$$z = \frac{\hat{p}_1 - \hat{p}_2}{\sqrt{\hat{p}(1-\hat{p})\left(\dfrac{1}{n_1} + \dfrac{1}{n_2}\right)}} = $$

Using Table A, compute

P-value =

Conclude: What do you conclude?

COMPLETE SOLUTIONS

Exercise 21.19

The count of successes and failures are each five or more in both samples, so the z test for equality of two population proportions can be used.

We are interested in determining whether there is good evidence that the proportion of papers rejected without review is *different* for papers with and without statistical help. Thus, the hypotheses to be tested are

$$H_0 \colon p_1 = p_2$$

$$H_a \colon p_1 \neq p_2$$

Letting population 1 be papers without statistical help and population 2 be papers with statistical help, the two sample sizes and estimates of the proportions are

$$n_1 = 190 \qquad \hat{p}_1 = 135/190 = 0.7105$$

$$n_2 = 514 \qquad \hat{p}_2 = 293/514 = 0.5700$$

The pooled sample proportion is

$$\hat{p} = \frac{\text{count of successes in both samples combined}}{\text{count of observations in both samples combined}} = \frac{135 + 293}{190 + 514} = \frac{428}{704} = 0.6080$$

and the z statistic is

$$z = \frac{\hat{p}_1 - \hat{p}_2}{\sqrt{\hat{p}(1 - \hat{p})\left(\dfrac{1}{n_1} + \dfrac{1}{n_2}\right)}} = \frac{0.7105 - 0.5700}{\sqrt{0.6080(1 - 0.6080)\left(\dfrac{1}{190} + \dfrac{1}{514}\right)}} = \frac{0.1405}{0.0414} = 3.39$$

Finally, using Table A, we have

$$P\text{-value} = 2 \times P(z \geq 3.39) = (2)(1 - 0.9997) = 0.0006$$

which is very strong evidence that a higher proportion of papers submitted without statistical help will be sent back without review. However, since this is an observational study, the evidence does not establish causation — getting statistical help for your paper may not make it less likely to be sent back without review.

Exercise 21.21

The counts of successes and failures are each 10 or more in both samples, so we may use the large-sample confidence interval.

Letting Population 1 be papers without statistical help and Population 2 be papers with statistical help, a success correspond to a paper being sent back without review, the two sample sizes, number of successes and estimates of the proportions

Population 1 $n_1 = 190$ number of successes = 135 $\hat{p}_1 = 135/190 = 0.7105$

Population 2 $n_2 = 514$ number of successes = 293 $\hat{p}_2 = 293/514 = 0.5700$

Using these values of $\hat{p}_1$ and $\hat{p}_2$ and these sample sizes, the standard error is

$$\text{SE} = \sqrt{\frac{\hat{p}_1(1 - \hat{p}_1)}{n_1} + \frac{\hat{p}_2(1 - \hat{p}_2)}{n_2}} = \sqrt{\frac{0.7105(1 - 0.7105)}{190} + \frac{0.5700(1 - 0.5700)}{514}} = \sqrt{0.001559} = 0.0395$$

For a 95% confidence interval, $z^* = 1.96$ and the interval is

$$(\hat{p}_1 - \hat{p}_2) \pm z^*\text{SE} = (0.7105 - 0.5700) \pm (1.96)(0.0395) = 0.1405 \pm 0.0774 = (0.0631, 0.2179)$$

We are 95% confident that the percentage of papers without statistical help that are not reviewed is between 6.3 and 21.8 percentage points higher than the percentage of papers with statistical help.

Exercise 21.29

State: What is the size of the difference in the proportion of mice ready to breed in good acorn years and bad acorn years? In a low-acorn year, experimenters added hundreds of thousands of acorns to area 1 to simulate a good acorn year, while area 2 was left untouched. They then trapped mice in both areas and counted the number of mice in breeding condition. The data follow.

Population	Population description	Sample size	Number of successes	Sample proportion
1	First area	$n_1 = 72$	54	$\hat{p}_1 = 54/72 = 0.75$
2	Second area	$n_2 = 17$	10	$\hat{p}_2 = 10/17 = 0.59$

Formulate: Give a 90% confidence interval for the difference of population proportions $p_1 - p_2$.

Solve: We can use the large-sample confidence interval when the populations are at least 10 times as large as the samples and the counts of successes and failures are 10 or more in both samples. For the 17 mice trapped in the second area, there are 10 successes and 7 failures. Because the number of failures in the second sample is less than 10, the large-sample confidence interval may not be accurate. Add four imaginary observations. The new data summary follows.

Population	Population description	Sample size	Number of successes	Plus four sample proportion
1	First area	$n_1 + 2 = 74$	$54 + 1 = 55$	$\tilde{p}_1 = 55/74 = 0.74$
2	Second area	$n_2 + 2 = 19$	$10 + 1 = 11$	$\tilde{p}_2 = 11/19 = 0.58$

The standard error is

$$SE = \sqrt{\frac{\tilde{p}_1(1-\tilde{p}_1)}{n_1} + \frac{\tilde{p}_2(1-\tilde{p}_2)}{n_2}} = \sqrt{\frac{0.74(1-0.74)}{74} + \frac{0.58(1-0.58)}{19}} = \sqrt{0.0026 + 0.0128} = 0.12$$

and for a 90% confidence interval $z^* = 1.645$ so the 90% confidence interval is

$$(\tilde{p}_1 - \tilde{p}_2) \pm z^*SE = (0.74 - 0.58) \pm 1.645(0.12) = 0.16 \pm 0.20$$

Conclude: We are 90% confident that mice in the first area (with abundant crop) are between –4% and 36% more likely to be in breeding condition than those in the second area, which was left untouched. The confidence interval is quite wide and there isn't much evidence of a difference in the breeding condition of mice in the two areas in either direction.

Exercise 21.32

State: North Carolina University looked at factors that affected the success of students in a required chemical engineering course. Students must receive a C or better in the course to continue as chemical engineering majors, so we consider a grade of C or better as a success. Is there a difference in the proportions of male and female students who succeeded in the course? The data showed that 23 of the 34 women and 60 of the 89 men succeeded. We view these as SRSs of men and women who would take this course.

Formulate: Let p_1 represent the proportion of female students who will succeed (population 1) and p_2 the proportion of males who will succeed (population 2). We are interested in determining whether there is a difference in these two proportions; hence we should test the hypotheses

$$H_0: p_1 = p_2$$

$$H_a: p_1 \neq p_2$$

Solve: We can use the z test when the count of successes and failures and successes are each 5 or more in both samples. For the females there were 23 successes and $34 - 23 = 11$ failures, and for the men there were 60 successes and $89 - 60 = 29$ failures. The conditions for safely using the test are met.

The two sample sizes are

$\quad n_1$ = number of female students in the course = 34

$\quad n_2$ = number of male students in the course = 89

From the data, the estimates of these two proportions are

$\quad \hat{p}_1 = 23/34 = 0.6765$

$\quad \hat{p}_2 = 60/89 = 0.6742$

We compute

$$\hat{p} = \frac{\text{count of successes in both samples combined}}{\text{count of observations in both samples combined}} = \frac{23+60}{34+89} = 83/123 = 0.6748$$

The value of the z-test statistic is thus

$$z = \frac{\hat{p}_1 - \hat{p}_2}{\sqrt{\hat{p}(1-\hat{p})\left(\frac{1}{n_1}+\frac{1}{n_2}\right)}} = \frac{0.6765 - 0.6742}{\sqrt{0.6748(1-0.6748)\left(\frac{1}{34}+\frac{1}{89}\right)}} = \frac{0.0023}{\sqrt{0.00892}} = 0.02$$

We compute the P-value using Table A (we need to double the tail area since this is a two-sided test):

$\quad P$-value $= 2 \times (0.4920) = 0.9840$

Conclude: The data provide no statistical evidence of a difference between the proportion of men and women who succeed.

CHAPTER 22

INFERENCE ABOUT VARIABLES: PART III REVIEW

To assist you in reviewing the material in Chapters 18 – 21, we provide the text chapter and related problems in this Study Guide for each of the odd numbered review exercises. Other than pointing you in the right direction, we provide no additional hints or solutions. At this point, you should be able to work these problems on your own with minimal assistance. As a final challenge, we encourage you to work some of the Supplementary Exercises, which integrate more fully the material in these chapters.

Exercise 22.1
Text Location – Chapter 19 for two-sample problems
Related Study Guide exercise – Exercise 19.4

Exercise 22.3
Text Location – Chapter 18 for one-sample t confidence interval
Related Study Guide exercise – Exercise 18.7

Exercise 22.5
(a) Text Location – Chapter 21 for testing equality of two proportions
Related Study Guide exercises – Exercises 21.9, 21.32

(b) Text Location – Chapter 19 for two-sample problems
Related Study Guide exercise – Exercise 19.4

(c) Text Location – Chapter 21 for testing equality of two proportions
Related Study Guide exercises – Exercises 21.9, 21.32

Exercise 22.7
Text Location – Chapter 9 for comparative experiments
Related Study Guide exercise – Exercise 9.6

Exercise 22.9
Text Location – Chapter 21 for testing equality of two proportions
Related Study Guide exercises – Exercises 21.9, 21.32

Exercise 22.11
Text Location – Chapter 19 for two-sample t procedures
Related Study Guide exercise – Exercise 19.39

Exercise 22.13
Text Location – Chapter 20 for a confidence interval about a proportion
Related Study Guide exercise – Exercise 20.9

Exercise 22.15
Text Location – Chapter 18 for one-sample t confidence interval
Related Study Guide exercise – Exercise 18.7

Exercise 22.17
Text Location – Chapter 20 for confidence interval about a proportion
Related Study Guide exercise – Exercise 20.9

Exercise 22.19
(a) Text Location – Chapter 19 for two-sample t procedures
Related Study Guide exercise – Exercise 19.34

(b) Text Location – Chapter 19 for robustness again
Related Study Guide exercise – Exercise 19.34

Exercise 22.21
Text Location – Chapter 21 for testing equality of two proportions
Related Study Guide exercises – Exercises 21.9, 21.32

Exercise 22.23
Text Location – Chapter 20 for confidence interval about a proportion
Related Study Guide exercise – Exercise 20.9

Exercise 22.25
Text Location – Chapter 20 for hypothesis test about a proportion
Related Study Guide exercise – Exercise 20.33

Exercise 22.27
Text Location – Chapter 19 for two-sample t procedures
Related Study Guide exercise – Exercise 19.34

Exercise 22.29
Text Location – Chapter 19 for two-sample t procedures
Related Study Guide exercise – Exercise 19.34

Exercise 22.31
Text Location – Chapter 21 for testing equality of two proportions
Related Study Guide exercises – Exercises 21.9, 21.32

Exercise 22.33
Text Location – Chapter 19 for two-sample t procedures
Related Study Guide exercise – Exercise 19.39

Exercise 22.35
Text Location – Chapter 18 for conditions for inference, Chapter 19 for comparing two population means
Related Study Guide exercises – Exercises 18.7, 19.34.

CHAPTER 23

TWO CATEGORICAL VARIABLES: THE CHI-SQUARE TEST

OVERVIEW

The inference methods of Chapter 20 of your text are first extended to a comparison of more than two population proportions. When comparing more than two proportions, it is best to first do an overall test to see if there is good evidence of any differences among the proportions. Then a detailed follow-up analysis can be performed to decide which proportions are different and to estimate the sizes of the differences.

The overall test for comparing several population proportions arranges the data in a **two-way table**. Two-way tables were introduced in Chapter 6 of your text; the tables are a way of displaying the relationship between any two categorical variables. The tables are also called $r \times c$ **tables**, where r is the number of rows and c is the number of columns. Often the rows of the two-way tables correspond to populations or treatment groups, the columns to different categories of the response. When comparing several proportions, there would be only two columns since the response takes only one of two values, and the two columns would represent the successes and failures.

The null hypothesis is $H_0 : p_1 = p_2 = \cdots = p_n$ which says that the n population proportions are the same. The alternative is "many sided," as the proportions can differ from each other in a variety of ways. To test this hypothesis, we will compare the **observed counts** with the **expected counts** when H_0 is true. The expected cell counts are computed using the formula

$$\text{expected count} = \frac{\text{row total} \times \text{column total}}{n}$$

where n is the total number of observations.

The statistic we will use to compare the expected counts with the observed counts is the **chi-square statistic**. It measures how far the observed and expected counts are from each other using the formula

$$X^2 = \sum \frac{(\text{observed} - \text{expected})^2}{\text{expected}}$$

where we sum up all the $r \times c$ cells in the table.

When the null hypothesis is true, the distribution of the test statistic X^2 is approximately the chi-squared distribution with $(r-1)(c-1)$ degrees of freedom. The P-value is the area to the right of X^2 under the chi-square density curve. Use Table E in the Appendix of the book to get the critical values and to compute the P-value. The mean of any chi-square distribution is equal to its degrees of freedom.

We can use the chi-square statistic when the data satisfy the following conditions.

- The data are independent SRSs from several populations and each observation is classified according to one categorical variable.

- The data are from a single SRS and each observation is classified according to two categorical variables.

- No more than 20% of the cells in the two-way table have expected counts less than 5.

- All cells have an expected count of at least 1.

- In the special case of the 2 × 2 table, all expected counts should exceed 5 before applying the approximation.

GUIDED SOLUTIONS

Exercise 23.1

KEY CONCEPTS: Conditional distributions in two-way tables

(a) The data from Exercise 23.1 are reproduced here to help you carry out the calculations. We have included a Total column that gives the row totals, which are required to obtain the conditional distributions.

Smoking Status

Education	Nonsmoker	Former	Moderate	Heavy	Total
Primary school	56	54	41	36	187
Secondary school	37	43	27	32	139
University	53	28	36	16	133

You are asked to first find the percent of men with a primary school education who are nonsmokers. There are a total of 187 men with a primary school education, and of these 56 are nonsmokers. Thus the percent of men with a primary school education who are nonsmokers is 56/187 = 0.30. Now compute the percent of men with a primary school education who are former smokers, moderate smokers, and heavy smokers. The total of the percents should add to 100%, except for possible roundoff error.

Smoking Status – Conditional Distributions

Education	Nonsmoker	Former	Moderate	Heavy	Total
Primary school	0.30				
Secondary school					
University					

(b) Complete the table in part (a) to give the conditional distributions of falling in each of the smoking categories for men with secondary school educations and university educations.

(c) Compare the three conditional distributions. Are there any clear relationships between smoking and education? How would you describe these relationships?

Exercise 23.5

KEY CONCEPTS: Computing and interpreting expected counts in a two-way table

(a) The data from Exercise 23.1 are reproduced here to help you carry out the calculations. We have included a Total column that gives the row totals and a Total row that gives the column totals, which are required to obtain the expected counts.

Smoking Status

Education	Nonsmoker	Former	Moderate	Heavy	Total
Primary school	56	54	41	36	187
Secondary school	37	43	27	32	139
University	53	28	36	16	133
Total	146	125	104	84	459

Remember that the expected cell counts are computed using the formula

$$\text{expected count} = \frac{\text{row total} \times \text{column total}}{n}$$

where n is the total number of observations. The value in the University row for Nonsmoker was obtained as

$$\text{expected count} = \frac{\text{row total} \times \text{column total}}{n} = \frac{(133)(146)}{459} = 42.30$$

Expected Counts for University Row

Education	Nonsmoker	Former	Moderate	Heavy	Total
University	42.30				

Fill in the rest of the expected counts. Verify that the row total of expected counts agrees with the row total of observed counts.

(b) Which cells have the largest deviations between expected and observed counts? What kind of relationship do these suggest?

Exercise 23.7

KEY CONCEPTS: Computing the X^2 statistic

(a) The contributions to the chi-square statistic are calculated from the expected and observed count in each cell. If you use the expected counts in the table to compute the components of the chi-square statistic, your results will be quite close to but not exactly equal to the values in the Minitab output. This is because the printed output rounds off the expected counts to two decimal places, but the calculations used by Minitab to compute the components of the chi-square statistic are based on "unrounded" values. So if your calculations are based on the "rounded-off" values, they may not exactly agree with the values in the Minitab output. Verify the contributions to chi-square for the University row. We have verified the entry for the Nonsmoker category.

Nonsmoker: $\dfrac{(\text{observed count} - \text{expected count})^2}{\text{expected count}} = \dfrac{(28 - 36.22)^2}{36.22} = 1.8655$

Former:

Moderate:

Heavy:

(b) The terms contributing most to X^2 are in the University row. Write a brief summary of the nature and significance of the relationship between education and smoking.

Exercise 23.9

KEY CONCEPTS: Cell counts required for chi-square test

You can safely use the chi-square test when no more than 20% of the expected counts are less than 5 and all expected counts are 1 or greater. Using the information in the Minitab output in Figure 23.3, verify that the data meet the requirement for use of chi-square.

Exercise 23.13

KEY CONCEPTS: Degrees of freedom for the chi-square distribution, P-values using Table E

(a) What are the values of r and c in the table? The degrees of freedom can be found using the formula

degrees of freedom $= (r-1)(c-1) =$

(b) Go to the row in Table E in the Appendix of the textbook corresponding to the degrees of freedom found in part (a). Between what two entries in Table E does the value $X^2 = 13.305$ lie? What does this tell you about the P-value?

Exercise 23.17

KEY CONCEPTS: Chi-square test for goodness of fit

(a) For the professor's grades, fill in the percentages of students earning each grade and compare them to the TA's percentages.

Grade	A	B	C	D/F
Percentage				

(b) Fill in the expected counts for each grade. Using the TA percentages as the null probabilities, the expected count for A's is $np_{10} = 91 \times 0.32 = 29.12$.

Grade	A	B	C	D/F
Expected	29.12			

(c) Complete the calculation of the chi-square goodness of fit statistic. We have indicated how the first term in the sum should be computed.

$$X^2 = \sum \frac{(\text{count of outcome } i - np_{i0})^2}{np_{i0}} = \frac{(22 - 29.12)^2}{29.12} +$$

The degrees of freedom are

$k - 1 =$

with k the number of possible outcomes. What can you say about the P-value and what is your conclusion?

Exercise 23.42

KEY CONCEPTS: Two-way tables, testing hypotheses with the X^2 statistic

(a) Think of the rows of the table (disease status) as the "treatments" and the columns (olive oil consumption) as the response; this designation corresponds to how the samples were selected. To be an experiment, what needs to be true about the assignment of the treatments to the subjects? Was this carried out here?

(b) If less than 4% of the cases or controls refused to participate, why would our confidence in the results be strengthened?

(c) The data from the text are reproduced here, with the expected counts printed below the observed counts. Verify a few of the expected counts on your own. If you are not sure how to compute the expected counts, review the complete solution for Exercise 23.7 of this study guide.

Olive Oil

	Low	Medium	High	Total
Colon cancer	398	397	430	1225
	404.39	404.19	416.42	
Rectal cancer	250	241	237	728
	240.32	240.20	247.47	
Controls	1368	1377	1409	4154
	1371.29	1370.61	1412.10	
Total	2016	2015	2076	6107

Is high olive oil consumption more common among patients without cancer than in patients with colon cancer or rectal cancer? Compute the percentages for each of the three below.

Group	Percentage with high olive oil consumption
Colon cancer	

Rectal cancer

Controls

What do these percentages suggest?

To do a formal test,

$$X^2 = \sum \frac{(\text{observed} - \text{expected})^2}{\text{expected}} =$$

What is the mean of the X^2 statistic under the null hypothesis? If there is evidence to reject the null hypothesis, the computed value of the statistic should be larger than we would expect it to be under the null hypothesis. Is that true in this case?

What is the P-value? What do you conclude?

COMPLETE SOLUTIONS

Exercise 23.1

(a), (b) The three conditional distributions are given here. Each entry is obtained by dividing the count in the cell by the row total. For example, the percent of men with a secondary school education who are moderate smokers is calculated as follows. There are a total of 139 men with a secondary school education. Of these, 27 are moderate smokers, so the percent of men with a secondary school education who are moderate smokers is 27/139 = 0.19.

Smoking Status – Conditional Distributions

Education	Nonsmoker	Former	Moderate	Heavy	Total
Primary school	0.30	0.29	0.22	0.19	1.00
Secondary school	0.27	0.31	0.19	0.23	1.00
University	0.40	0.21	0.27	0.12	1.00

(c) There are several things to note. The first is that the conditional distribution of smoking status is almost the same for men with a primary school education and men with a secondary school education. The conditional distribution for men with a university education differs from the other two in several clear ways. The percent in the moderate smoking status plus the percent in the heavy smoking status is about the same for primary school, 0.22 + 0.19 = 0.41, secondary school, 0.19 + 0.23 = 0.42, and university 0.27 + 0.12 = 0.39. However, among those that smoke fewer are heavy smokers in the university education level.

The percent in the nonsmoker plus the percent in the former smoker category is also about the same for all education levels (it has to be since the percents add to 100%). However, there is a much higher percentage of men with a university education who have never smoked; 40% versus 30% and 27% in the primary school and secondary school level of education, respectively. Ultimately there are about the same percent of men who do not smoke in all three educational categories, but the university educated tended to be more likely not to start in the first place.

Exercise 23.5

(a) Following is the table of expected counts with the required calculations.

Expected Counts for the University Row

Education	Nonsmoker	Former	Moderate	Heavy	Total
University	42.30	36.22	30.14	24.33	132.99

$$36.22 = \frac{(133)(125)}{459} \qquad 30.14 = \frac{(133)(104)}{459} \qquad 24.34 = \frac{(133)(84)}{459}$$

The total of the expected counts in this row is 132.99 which agrees with the total of observed counts except for roundoff error.

(b) There are more nonsmokers than the null hypothesis calls for. Also, there are fewer former smokers, more moderate smokers and fewer heavy smokers. See the discussion in part (c) of Exercise 23.1 for a better understanding of this result.   It is true that university-educated men smoke less than the null hypothesis calls for, but this is a slight oversimplification of the data. More university-educated men have never smoked, and among current smokers they smoke less (more moderate smokers and fewer heavy smokers).

Exercise 23.7

(a) The components of the chi-square statistic are calculated from the expected and observed count in each cell for the University row.

Nonsmoker: $\dfrac{(\text{observed count} - \text{expected count})^2}{\text{expected count}} = \dfrac{(28 - 36.22)^2}{36.22} = 1.8655$

Former: $\dfrac{(\text{observed count} - \text{expected count})^2}{\text{expected count}} = \dfrac{(16 - 24.34)^2}{24.34} = 2.8577$

Moderate: $\dfrac{(\text{observed count} - \text{expected count})^2}{\text{expected count}} = \dfrac{(36 - 30.14)^2}{30.14} = 1.1393$

Heavy: $\dfrac{(\text{observed count} - \text{expected count})^2}{\text{expected count}} = \dfrac{(53 - 42.31)^2}{42.31} = 2.7009$

(b) The terms contributing most to X^2 are in the University row. There are more nonsmokers than the null hypothesis calls for. Also, there are fewer former smokers, more moderate smokers, and fewer heavy smokers. Generally, university-educated men smoke less than the null hypothesis calls for. More university-educated men have never smoked, and among current smokers they smoke less (more moderate smokers and fewer heavy smokers). See the discussion in part (c) of Exercise 23.1

Exercise 23.9

Using the information in the Minitab output in Figure 23.3, we see that all 12 of the expected counts exceed 5. The smallest expected count is 24.34, so the data easily meet the cell requirements for safe use of chi-square.

Exercise 23.13

(a) There are $r = 3$ rows and $c = 4$ columns. Thus the number of degrees of freedom is

$$(r-1)(c-1) = (3-1)(4-1) = (2)(3) = 6$$

which agrees with the value in the Minitab output.

(b) If we look in the df $= 6$ row of Table E, we find the following information.

p	.05	.025
x^*	12.59	14.45

$X^2 = 13.305$ lies between the entries for $p = .05$ and $p = .025$. This tells us that the P-value is between .05 and .025.

Exercise 23.17

(a) For the professor's grades, fill in the percentages of students earning each grade and compare them to the TA's percentages.

Grade	A	B	C	D/F
Percentage	22/91 = 0.24	38/91 = 0.42	20/91 = 0.22	11/91 = .0.12

The professor gave a smaller percentage of A's and a higher percentage of D's and F's than the TA. The percentages of B's and C's were similar.

(b) Using the TA percentages as the null probabilities, the expected counts for the professor are

$np_{10} = 91 \times 0.32 = 29.12$

$np_{20} = 91 \times 0.41 = 37.31$

$np_{30} = 91 \times 0.20 = 18.20$

$np_{40} = 91 \times 0.07 = 6.37$

Grade	A	B	C	D/F
Expected	29.12	37.31	18.20	6.37

(c) The value of the test statistic is

$$X^2 = \frac{(22-29.12)^2}{29.12} + \frac{(38-37.31)^2}{37.31} + \frac{(20-18.20)^2}{18.20} + \frac{(11-6.37)^2}{6.37} = 1.741 + 0.022 + 0.178 + 3.365 = 5.306$$

The degrees of freedom are $k - 1 = 4 - 1 = 3$. Looking in the df = 3 row of Table E, we find the information

p	.20	.15
x^*	4.64	5.32

giving a P-value between 0.15 and 0.20. There is little evidence that the professor's grade distribution differs from the TA's.

Exercise 23.42

(a) We are thinking of the disease groups as the explanatory variable, since the samples were chosen from these groups. The experimenters did not assign the subjects to the disease groups. This makes the data observational, and there is the possibility that differences in olive oil consumption may be a result of confounding variables, not necessarily disease status. (It might be more natural to think of olive oil consumption as the explanatory variable and disease status as the response, but the samples were not selected from the different olive oil consumption groups.)

(b) If there is a high *nonresponse rate*, then it is possible for the percentages in the table to be quite different than the true percentages due to response bias. In this case, the existence of or absence of a relationship could potentially be due to this bias. However, with 96% of the people responding, the bias if it exists could not be large and would have little impact on the results.

(c) The percentages in each group with high olive oil consumptions are computed in the table. There appears to be very little difference in these percentages, suggesting that olive oil in the diet may be unrelated to these forms of cancer.

Group	Percentage with high olive oil consumption
Colon cancer	430/1225 = 0.351
Rectal cancer	237/728 = 0.326
Controls	1409/4154 = 0.339

Using the components of chi-square from Minitab

$$X^2 = \sum \frac{(\text{observed} - \text{expected})^2}{\text{expected}} = 0.101 + 0.128 + 0.443 + 0.390 + 0.003 + 0.443 +$$

$$0.008 + 0.030 + 0.007 = 1.552$$

If you used the expected counts in the table, which were rounded to two decimals, your answers may disagree slightly with the computations due to rounding error. The mean of the X^2 statistic is equal to the degrees of freedom, which in this case is $(r - 1)(c - 1) = (3 - 1)(3 - 1) = 4$. Since the numerical value of the X^2 statistic is below the mean, there is little evidence to reject the null hypothesis. As can be seen from the table, the observed and expected counts are quite close. From Table E, you can see that the P-value is greater than 0.25. In fact, using statistical software shows that the P-value = 0.817. These data show no evidence of a relationship between disease and olive oil consumption.

CHAPTER 24

INFERENCE FOR REGRESSION

OVERVIEW

In Chapter 5 of the textbook we first encountered regression. The assumptions that describe the regression model we use in this chapter are the following.

- We have n observations on an explanatory variable x and a response variable y. Our goal is to study or predict the behavior of y for given values of x.

- For any fixed value of x, the response y varies according to a normal distribution. Repeated responses y are independent of each other.

- The mean response μ_y has a straight-line relationship with x:

$$\mu_y = \alpha + \beta x$$

The slope β and intercept α are unknown parameters.

- The standard deviation of y (call it σ) is the same for all values of x. The value of σ is unknown.

The **true regression line** is $\mu_y = \alpha + \beta x$ and says that the mean response μ_y moves along a straight line as the explanatory variable x changes. The parameters β and α are estimated by the slope b and intercept a of the least-squares regression line, and the formulas for these estimates are

$$b = r \frac{s_y}{s_x}$$

and

$$a = \bar{y} - b\,\bar{x}$$

where r is the correlation between y and x, $\bar{y}$ is the mean of the y observations, s_y is the standard deviation of the y observations, $\bar{x}$ is the mean of the x observations, and s_x is the standard deviation of the x observations.

The **standard error about the least-squares line** is

$$s = \sqrt{\frac{1}{n-2}\sum \text{residual}^2} = \sqrt{\frac{1}{n-2}\sum (y - \hat{y})^2}$$

where $\hat{y} = a + bx$ is the value we would predict for the response variable based on the least-squares regression line. We use s to estimate the unknown σ in the regression model.

A **level C confidence interval** for β is

$$b \pm t^* SE_b$$

where t^* is the upper $(1 - C)/2$ critical value for the t distribution with $n - 2$ degrees of freedom and

$$SE_b = \frac{s}{\sqrt{\Sigma(x - \bar{x})^2}}$$

is the standard error of the least-squares slope b. SE_b is usually computed using a calculator or statistical software.

The **test of the hypothesis** H_0: $\beta = 0$ is based on the t statistic

$$t = \frac{b}{SE_b}$$

with P-values computed from the t distribution with $n - 2$ degrees of freedom. This test is also a test of the hypothesis that the correlation is 0 in the population.

A **level C confidence interval for the mean response** μ_y when x takes the value x^* is

$$\hat{y} \pm t^* SE_{\hat{\mu}}$$

where $\hat{y} = a + bx$, t^* is the upper $(1 - C)/2$ critical value for the t distribution with $n - 2$ degrees of freedom, and

$$SE_{\hat{\mu}} = s\sqrt{\frac{1}{n} + \frac{(x^* - \bar{x})^2}{\Sigma(x - \bar{x})^2}}$$

$SE_{\hat{\mu}}$ is usually computed using a calculator or statistical software.

A **level C prediction interval for a single observation** on y when x takes the value x^* is

$$\hat{y} \pm t^* SE_{\hat{y}}$$

where t^* is the upper $(1 - C)/2$ critical value for the t distribution with $n - 2$ degrees of freedom and

$$SE_{\hat{y}} = s\sqrt{1 + \frac{1}{n} + \frac{(x^* - \bar{x})^2}{\Sigma(x - \bar{x})^2}}$$

$SE_{\hat{y}}$ is usually computed using a calculator or statistical software.

Finally, it is always good practice to check that the data satisfy the linear regression model assumptions before doing inference. Scatterplots and residual plots are useful tools for checking these assumptions.

GUIDED SOLUTIONS

Exercise 24.1

KEY CONCEPTS: Scatterplots, correlation, linear regression, residuals, standard error of the least-squares line

(a) Sketch your scatterplot on the axes provided.

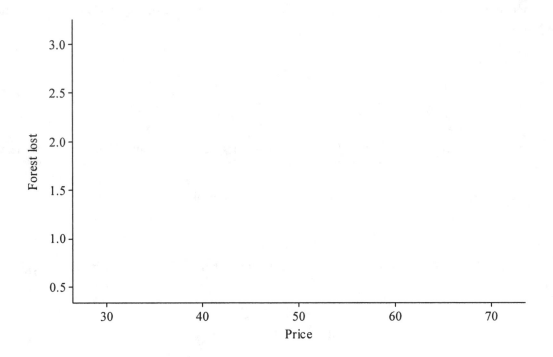

Use your calculator (or statistical software) to compute the correlation r and the equation of the least-squares regression line.

   $r =$

   Forest lost $=$

Do you think coffee price will allow a good prediction of forest lost?

(b) What does the slope β of the true regression line say about coffee price and forest lost?

Enter your estimates of the slope β and intercept α of the true regression line. Refer to your answer in part (a) for these estimates.

Estimate of $\beta =$

Estimate of $\alpha =$

(c) To compute the residuals, complete the table.

Observed value of forest lost	Predicted value of forest lost : $-1.0134 + 0.05525(\text{price})$	Residual (observed – predicted)
0.49		
1.59		
1.69		
1.82		
3.10		

Sum =

Now estimate the standard deviation σ by computing

$$\sum \text{residual}^2 \ =$$

and then completing the following calculation.

$$s = \sqrt{\frac{1}{n-2} \sum \text{residual}^2} \ =$$

Exercise 24.4

KEY CONCEPTS: Tests for the slope of the least-squares regression line

(a)The test of the hypothesis H_0: $\beta = 0$ is based on the t statistic

$$t = \frac{b}{\text{SE}_b}$$

In the statement of the problem, we are told that $b = 0.0543$ and $\text{SE}_b = 0.0097$. The value of b is slightly different than the value we found in Exercise 24.1, but we will use it to compute t.

$$t = \frac{b}{\text{SE}_b} \ =$$

(b) What are the degrees of freedom for t? Refer to the original data in Exercise 24.1 of your textbook to determine the sample size n.

Degrees of freedom $= n - 2 =$

Now use Table C to estimate the P-value.

P-value:

What do you conclude?

Exercise 24.41

KEY CONCEPTS: Scatterplots, examining residuals, confidence intervals for the slope

The four-step process follows.

State. What is the practical question in the context of the real-world setting?

Formulate. What specific statistical operations does this problem call for?

Solve. Make the graphs and carry out any calculations needed for this problem.

Conclude. Give your practical conclusion in the setting of the real-world problem.

To apply the steps to this problem, here are some suggestions.

State. What characteristics of the young children are of interest here? What question about these characteristics do we wish to answer?

Formulate. What inferences are we asked to make?

Solve. Answering parts (a), (b), and (c) will complete the solve step.

(a) Use software or a calculator to compute the equation of the least-squares regression line.

$\hat{y} =$

Use software or the axes provided to make your scatterplot.

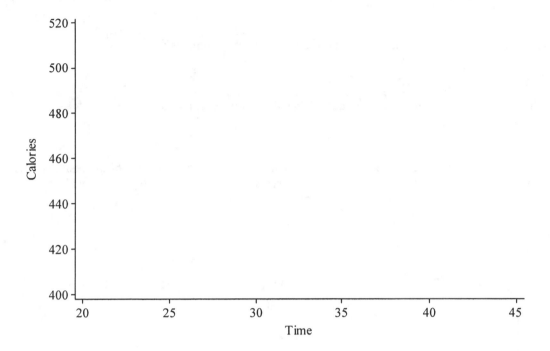

(b) Use software or the axes provided to make your plot of the residuals.

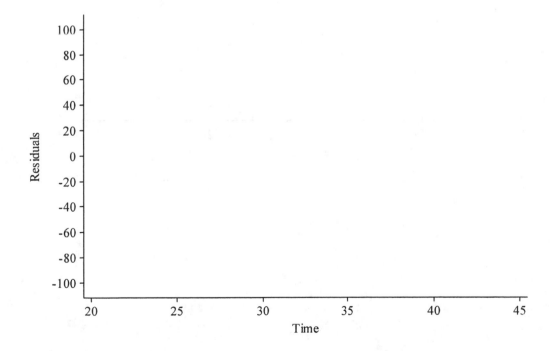

Are the observations independent? Why?

Does the relationship appear linear? What does your scatterplot indicate?

Does the spread about the line stay the same? What does your residual plot suggest?

Does the variation about the line appear to be normal? What does a histogram of the residuals suggest? Use software or the axes that follow (with class intervals $-40 \leq$ residual < -30, $-30 \leq$ residual < -20, $-20 \leq$ residual < -10, and so on) to make a histogram.

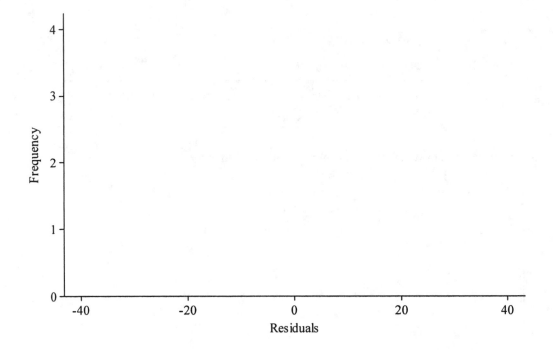

(c) To determine the confidence interval, recall that a level C confidence interval for β is

$$b \pm t^*SE_b$$

where t^* is the upper $(1 - C)/2$ critical value for the t distribution with $n - 2$ degrees of freedom and

$$SE_b = \frac{s}{\sqrt{\sum (x - \bar{x})^2}}$$

is the standard error of the least-squares slope b. In this exercise, b and SE_b can be read directly from the output of statistical software. Record their values.

$b =$

$SE_b =$

Now find t^* for a 90% confidence interval from Table C (what is n here?).

$t^* =$

Put all these pieces together to compute the 90% confidence interval.

$b \pm t^* SE_b =$

Conclude. What do the data show about the behavior of children?

Exercise 24.43

KEY CONCEPTS: Prediction, prediction intervals

We ran Minitab and asked for prediction at Time = 40. The output follows.

```
The regression equation is
Calories = 561 - 3.08 Time

Predictor        Coef        Stdev      t-ratio          p
Constant       560.65        29.37        19.09      0.000
Time          -3.0771        0.8498       -3.62      0.002

s = 23.40        R-sq = 42.1%      R-sq(adj) = 38.9%

Analysis of Variance

SOURCE          DF            SS           MS          F          p
Regression       1        7177.6       7177.6      13.11      0.002
Error           18        9854.4        547.5
Total           19       17032.0

     Fit   Stdev.Fit         95.0% C.I.           95.0% P.I.
  437.57        7.30    ( 422.23,   452.91)   ( 386.06,   489.08)
```

Where in this output does one find the 95% confidence interval to predict Rachel's calorie consumption at lunch? Refer to Examples 24.7 and 24.8 in the textbook if you need help.

95% prediction interval:

COMPLETE SOLUTIONS

Exercise 24.1

(a) If we look at the data, we see that as the coffee prices increase, so does the percent of forest area lost. Thus there is a positive association between price and forest lost. A scatterplot of the data with price as the explanatory variable follows.

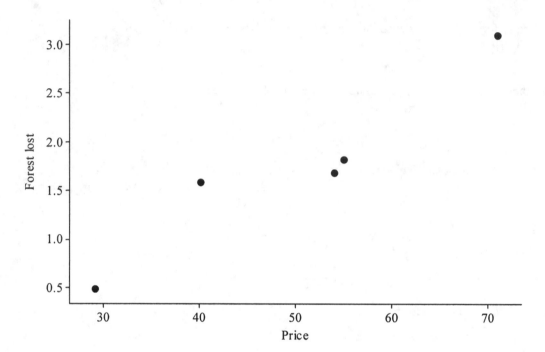

The scatterplot indicates a fairly strong positive association between price and forest lost. If we calculate the correlation r and the equation of the least-squares line, we obtain the following.

$$r = 0.952$$

$$\text{Forest lost} = -1.0134 + 0.05525 \text{ (price)}$$

The correlation is high, so one would expect that price would allow reasonably good prediction of forest lost.

(b) The slope β of the true regression line tells us the mean increase (percent) in forest area lost associated with a 1-cent increase in the price paid to coffee growers. From the data,

$$\text{Estimate of } \beta = 0.05525$$

the slope of the least-squares regression line. From the data,

$$\text{Estimate of } \alpha = -1.0134$$

the intercept of the least-squares regression line.

(c) The residuals for the five data points are given in the table.

Observed value of forest lost	Predicted value of forest lost : $-1.0134 + 0.05525(\text{price})$	Residual (observed – predicted)
0.49	0.58885	–0.09885
1.59	1.1966	0.3934
1.69	1.9701	–0.2801
1.82	2.02535	–0.20535
3.10	2.90935	0.19065

The sum of the residuals listed is –.00025, the difference from 0 due to roundoff. To estimate the standard deviation σ in the regression model, we first calculate the sum of the squares of the residuals listed:

$$\sum \text{residual}^2 = 0.321507$$

Our estimate of the standard deviation σ in the regression model is therefore

$$s = \sqrt{\frac{1}{n-2}\sum \text{residual}^2} = \sqrt{\frac{1}{5-2}(0.321507)} = 0.327367$$

Exercise 24.4

(a) $b = 0.0543$ and $SE_b = 0.0097$, so

$$t = \frac{b}{SE_b} = \frac{0.0543}{0.0097} = 5.6$$

(b) Referring to the original data in Exercise 24.4 of the textbook, we see that $n = 5$.

Degrees of freedom $= n - 2 = 5 - 2 = 3$

To estimate the P-value, we use Table C with df = 17 and find the two values that bracket the computed value of $t = 5.6$.

df = 17		
p	0.01	0.005
t^*	4.541	5.841

Because the test is one-sided, $0.005 < P$-value < 0.01. Statistical software gives a P-value of 0.005625 for $t = 5.6$.

We look in the row corresponding to 3 df. We see that the largest entry in the row is 12.92. Because our t is larger than this, we conclude

P-value < 0.0005

Exercise 24.41

State. We are interested in the time children remain at the lunch table and whether this can help us predict the calories they consume.

Formulate. We will make a scatterplot of the average number of minutes (time) a child spent at the table when lunch was served versus the average number of calories the child consumed. Time is the explanatory variable and calories the response. We will also compute the least-squares regression line to help us describe the relationship. We will check the conditions for inference. Finally, we will construct a 95% confidence interval to estimate the slope of the least-squares regression line. This slope tells us how rapidly calories consumed changes as time at the table increases.

Solve. (a) Here is a scatterplot showing the relationship between time at the table and calories consumed. Since we are trying to use time to explain calories, time is the explanatory variable and goes on the horizontal axis.

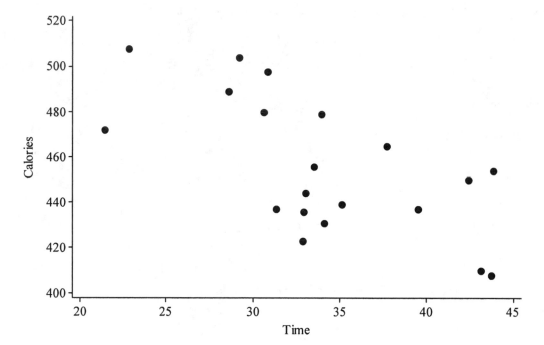

The overall pattern is roughly linear with a negative slope. There are no clear outliers or strongly influential data points.

Using statistical software, we find that the equation of the least-squares line is

$$\hat{y} = 560.65 - 3.08 \times time$$

(b) A scatterplot of the residuals follows.

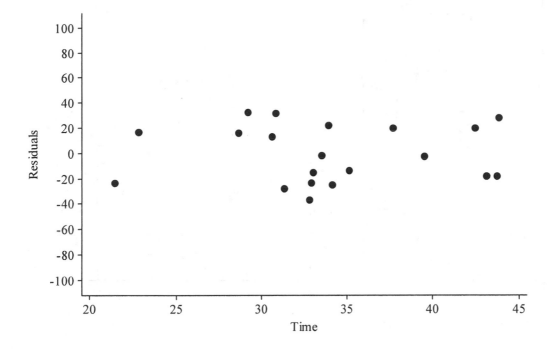

Are the observations independent?

The answer is not clear. These are observations on 20 different children rather than on a single child, and that is good. However, we do not know if the children were selected at random. In addition, we do not know if the children were all together so that the behavior of one child could influence the behavior of another.

Does the relationship appear linear?

The scatterplot shows a roughly linear trend sloping down from left to right.

Does the spread about the line stay the same?

The residual plot seems to suggest that the spread about the line is roughly constant. Points seem to lie consistently in a band between –40 and +40.

Does the variation about the line appear to be normal?

The histogram that follows has a gap and is not particularly bell-shaped. On the other hand there do not appear to be any outliers or extreme skewness.

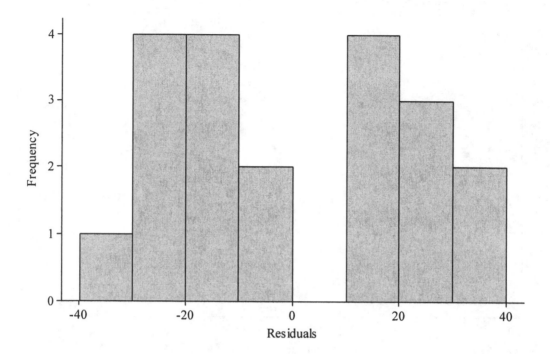

If we can regard the observations as independent, the conditions for inference (for a sample of size 20) are approximately satisfied.

(c) From statistical software we find that

$$b = -3.08$$

$$SE_b = 0.85$$

For a 95% confidence interval from Table C with $n = 20$ (and $n - 2 = 18$),

$$t^* = 2.101$$

We put these pieces together to compute the 95% confidence interval for the true slope of the regression line.

$$b \pm t^* SE_b = -3.08 \pm (2.101)(0.85) = -3.08 \pm 1.79$$

Conclude. The scatterplot shows that the more time a child spent at the table, the fewer the number of calories the child consumed. The least-squares regression line tells us that each minute increase in time spent at the table is associated with 3.08 fewer calories consumed. The 95% confidence interval does not contain 0 and suggests that the slope is clearly negative, confirming the pattern observed in the scatterplot and suggested by the least-squares regression line.

Exercise 24.43

The output from Minitab follows.

```
The regression equation is
Calories = 561 - 3.08 Time

Predictor        Coef        Stdev      t-ratio         p
Constant       560.65        29.37        19.09     0.000
Time          -3.0771       0.8498        -3.62     0.002

s = 23.40        R-sq = 42.1%      R-sq(adj) = 38.9%

Analysis of Variance

SOURCE          DF           SS          MS          F         p
Regression       1       7177.6      7177.6      13.11     0.002
Error           18       9854.4       547.5
Total           19      17032.0

      Fit    Stdev.Fit       95.0% C.I.             95.0% P.I.
   437.57         7.30    (422.23,  452.91)     (386.06,  489.08)
```

The "Fit" entry gives the predicted calories. Minitab gives both the 95% confidence interval for the mean response and the prediction interval for a single observation. We are predicting a single observation, so the column labeled "95% PI" contains the interval we want. We see

95% prediction interval: (386.06, 489.08)

CHAPTER 25

ONE-WAY ANALYSIS OF VARIANCE: COMPARING SEVERAL MEANS

OVERVIEW

The two-sample t procedures compare the means of two populations. However, when the mean is the best description of the center of a distribution, we may want to compare several population means or several treatment means in a designed experiment. For example, we might be interested in comparing the mean weight loss by dieters on three different diet programs or the mean yield of four varieties of green beans.

The method we use to compare more than two population means is the **analysis of variance (ANOVA) F test**. This test is also called the **one-way ANOVA**. The ANOVA F test is an overall test that looks for any difference between a group of I means. The null hypothesis is H_0: $\mu_1 = \mu_2 = \cdots = \mu_I$, where we tell the population means apart by using the subscripts 1 through I. The alternative hypothesis is H_a: not all the means are equal. In a more advanced course, you would study formal inference procedures for a follow-up analysis to decide which means differ and to estimate how large the differences are. Note that formally the ANOVA F test is a different test from the F test introduced in Chapter 19 of your text that compared the standard deviations of two populations, although the ANOVA F test does involve the comparison of two measures of variation.

The ANOVA F test compares the variation among the groups to the variation within the groups through the **F statistic**,

$$F = \frac{\text{variation among the sample means}}{\text{variation among the individuals in the same sample}}$$

The important thing to take away from this chapter is the rationale behind the ANOVA F test. The particulars of the calculation are not as important since software usually calculates the numbers for us.

The F statistic has the F distribution. The distribution is completely defined by its two degrees of freedom parameters, the numerator degrees of freedom and the denominator degrees of freedom. The numerator has $I - 1$ degrees of freedom, where I is the number of populations we are comparing. The denominator has $N - I$ degrees of freedom, where N is the total number of observations. The F distribution is usually written $F(I - 1, N - I)$.

We make the following assumptions for ANOVA:

- There are I independent SRSs.
- Each population is normally distributed with its own mean, μ_i.
- All populations have the same standard deviation, σ.

The first assumption is the most important. The test is robust against nonnormality, but it is still important to check for outliers and/or skewness that would make the mean a poor measure of the center of the distribution. As for the assumption of equal standard deviations, make sure that the largest sample standard deviation is no more than twice the smallest standard deviation.

Although it is generally best to leave the ANOVA computations to statistical software, seeing the formulas sometimes helps one to obtain a better understanding of the procedure. In addition, there are times when the original data are not available and you have only the group means and standard deviations or standard error. In these instances, the formulas described here are required to carry out the ANOVA F test.

The F statistic is $F = \dfrac{\text{MSG}}{\text{MSE}}$, where MSG is the **mean square for groups,**

$$\text{MSG} = \frac{n_1(\bar{x}_1 - \bar{x})^2 + n_2(\bar{x}_2 - \bar{x})^2 + \cdots + n_I(\bar{x}_I - \bar{x})^2}{I - 1}$$

with

$$\bar{x} = \frac{n_1\bar{x}_1 + n_2\bar{x}_2 + \cdots + n_I\bar{x}_I}{N}$$

and MSE is the **error mean square,**

$$\text{MSE} = \frac{s_1^2(n_1 - 1) + s_2^2(n_2 - 1) + \cdots + s_I^2(n_I - 1)}{N - I}.$$

Because MSE is an average of the individual sample variances, it is also called the **pooled sample variance,** written s_P^2, and its square root, $s_p = \sqrt{\text{MSE}}$ is called the **pooled standard deviation**. We can also make a confidence interval for any of the means by using the formula $\bar{x}_i \pm t^* \dfrac{s_p}{\sqrt{n_i}}$. The critical value is t^* from the t distribution with $N - I$ degrees of freedom.

GUIDED SOLUTIONS

Exercise 25.3

KEY CONCEPTS: Side-by-side stemplots, ANOVA hypotheses, drawing conclusions from ANOVA output

(a) Complete the stemplots on the next page (they use split stems). From the stemplots, would you say that any of the groups show outliers or extreme skewness? What effects of logging are visible from the stemplots?

Never logged	Logged 1 year ago	Logged 8 years ago
0	0	0
0	0	0
1	1	1
1	1	1
2	2	2
2	2	2
3	3	3

(b) What do the means suggest about the effect of logging?

(c) State the null and alternative hypotheses, letting μ_1, μ_2 and μ_3 denote the means for the three groups.

H_0: H_a:

From the output, determine the values of the ANOVA F statistic and its P-value. What are your conclusions?

F statistic = P-value =

Exercise 25.10

KEY CONCEPTS: ANOVA degrees of freedom, computing P-values from Table D

(a) In the table, fill in the numerical values and explain in words the meaning of each symbol we are using in the notation for the one-way ANOVA. Group 1, group 2, and group 3 are identified in the exercise.

Symbol	Value	Verbal meaning
I		
n_1		
n_2		
n_3		
N		

(b) Use the text formulas and the results from part (a) to give the numerator and denominator degrees of freedom. Check your answers against the Excel output given in Exercise 25.3.

Numerator degrees of freedom =

Denominator degrees of freedom =

(c) The value $F = 11.43$ needs to be referred to an $F(2, 30)$ distribution. What can you say about the P-value from Table D?

Exercise 25.13

KEY CONCEPTS: Checking standard deviations, ANOVA computations

(a) Do the standard deviations satisfy the rule of thumb for using ANOVA?

$$\frac{\text{largest sample standard deviation}}{\text{smallest sample standard deviation}} =$$

(b) You will need the means, sample sizes and standard deviations for the three groups to do the calculations. To compute MSG, you first need to compute the overall mean

$$\bar{x} = \frac{n_1\bar{x}_1 + n_2\bar{x}_2 + \cdots + n_I\bar{x}_I}{N} =$$

and then substitute the means, sample sizes, and overall mean into the formula

$$MSG = \frac{n_1(\bar{x}_1 - \bar{x})^2 + n_2(\bar{x}_2 - \bar{x})^2 + \cdots + n_I(\bar{x}_I - \bar{x})^2}{I - 1} =$$

(c) MSE is then obtained from the formula

$$MSE = \frac{s_1^2(n_1 - 1) + s_2^2(n_2 - 1) + \cdots + s_I^2(n_I - 1)}{N - I} =$$

(d) The F statistic is calculated as

$$F = \frac{MSG}{MSE} =$$

What are the degrees of freedom for the ANOVA F statistic? Compare the value you calculated to the critical values in Table D. What is the P-value and is there evidence that mean weight losses of people who follow the three exercise programs differ?

Exercise 25.35

KEY CONCEPTS: Four-step rule, one-way analysis of variance

The four-step process follows.

State: What is the practical question in the context of the real-world setting?

Formulate: What specific statistical operations does this problem call for?

Solve: Make the graphs and carry out any calculations needed for this problem.

Conclude: Give your practical conclusion in the setting of the real-world problem.

To apply the steps to this problem, here are some suggestions. You may want to use Example 25.4 of your text as a guide.

State: Using the language in the problem, state the goals of the study. Since the data set is small, you can include the data as well.

Formulate: Be sure to check to make sure that you can safely use ANOVA.

Solve: The output here is from MINITAB. If you are using software for your course, you should try to run the one-way ANOVA for this problem on your own software. Carrying out the calculations by hand is quite tedious even for a small data set such as this one.

One-way ANOVA: Breaking Strength versus Weeks

```
Source  DF      SS     MS     F      P
Weeks    2   381.7  190.9  3.70  0.056
Error   12   619.2   51.6
Total   14  1000.9

S = 7.183   R-Sq = 38.14%   R-Sq(adj) = 27.83%
```

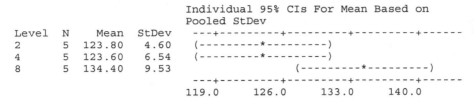

```
                          Individual 95% CIs For Mean Based on
                          Pooled StDev
Level  N    Mean   StDev   ---+---------+---------+---------+------
2      5  123.80   4.60    (---------*---------)
4      5  123.60   6.54    (---------*---------)
8      5  134.40   9.53                     (---------*---------)
                           ---+---------+---------+---------+------
                          119.0     126.0      133.0     140.0
```

Pooled StDev = 7.18

Conclude: Be careful with your conclusion. Even though the data shows some evidence of a difference in breaking strength over weeks, do the data support the conjecture that polyester loses strength over time?

COMPLETE SOLUTIONS

Exercise 25.3

(a) The side-by-side stemplots are completed here. Extreme skewness is not evident. There is a low outlier in the "Logged 8 years ago" column. The counts of trees in the plots that were never logged appear to be larger than those that were logged in the stemplots. There appears to be little difference between the counts from plots logged 1 year ago and plots logged 8 years ago.

Never logged	Logged 1 year ago	Logged 8 years ago

```
0  |             0 | 2         0 | 4
0  |             0 | 9         0 |
1  |             1 | 2 2 4 4   1 | 2 2
1  | 6 9 9       1 | 5 7 7 8 9 1 | 5 8 8 9
2  | 0 1 2 4     2 | 0         2 | 2 2
2  | 7 7 8 9     2 |           2 |
3  | 3           3 |           3 |
```

(b) The means suggest that logging reduces the number of trees per plot and that the recovery may be quite slow as there is little difference in the means for the logged 1 year ago and logged 8 years ago groups.

(c) The hypotheses are $H_0: \mu_1 = \mu_2 = \mu_3$ and H_a: not all of μ_1, μ_2 and μ_3 are equal.

The overall ANOVA F test has $F = 11.4257$ and P-value = 0.000205, so there is strong evidence of a difference in mean trees per plot among the three groups. The ANOVA F test does not tell us which groups are different, but examination of the means and stemplots shows both the logged groups to have lower mean trees per plot than the never logged group, but there appears to be little difference between the plots logged 1 year ago and the plots logged 8 years ago.

Exercise 25.10

(a) The table gives the value and states in words the meaning of each symbol used in a one-way ANOVA.

Symbol	Value	Verbal meaning
I	3	Number of groups
n_1	12	Number of plots in the never logged group
n_2	12	Number of plots in the logged 1 year ago group
n_3	9	Number of plots in the logged 8 years ago group
N	33	Total number of plots in the experiment

(b) The ANOVA F statistic has the F distribution with $I - 1 = 3 - 1 = 2$ degrees of freedom in the numerator and $N - I = 33 - 3 = 30$ degrees of freedom in the denominator.

(c) The critical value of 9.22 corresponds to a tail probability of 0.001 for an $F(2, 25)$ distribution. Since the critical value for an $F(2, 30)$ distribution corresponding to a tail area of 0.001 would be even smaller, the value $F = 11.4257$ would exceed this value. We can say that the P-value is less than 0.001, which agrees with the computer output.

Exercise 25.13

(a) The ratio of the largest to the smallest standard deviations is

$$\frac{\text{largest sample standard deviation}}{\text{smallest sample standard deviation}} = \frac{5.2}{4.2} = 1.24 < 2$$

so the rule of thumb for safe use of ANOVA is satisfied.

(b) We have three treatments, so $I = 3$ and the formula for the overall mean gives

$$\bar{x} = \frac{n_1\bar{x}_1 + n_2\bar{x}_2 + \cdots + n_I\bar{x}_I}{N} = \frac{n_1\bar{x}_1 + n_2\bar{x}_2 + \cdots + n_I\bar{x}_I}{n_1 + n_2 + n_3} = \frac{(37)(10.2)+(36)(9.3)+(42)(10.2)}{37+36+42} = \frac{1140.6}{115} = 9.9183$$

and then substituting the means, sample sizes, and overall mean into the formula

$$\text{MSG} = \frac{n_1(\bar{x}_1 - \bar{x})^2 + n_2(\bar{x}_2 - \bar{x})^2 + \cdots + n_I(\bar{x}_I - \bar{x})^2}{I - 1}$$

$$= \frac{37(10.2 - 9.9183)^2 + 36(9.3 - 9.9183)^2 + 42(10.2 - 9.9183)^2}{3 - 1} = \frac{20.0317}{2} = 10.016$$

(c) Remembering to square the standard deviations, MSE is then obtained from the formula

$$\text{MSE} = \frac{s_1^2(n_1 - 1) + s_2^2(n_2 - 1) + \cdots + s_I^2(n_I - 1)}{N - I}$$

$$= \frac{(4.2)^2(37 - 1)+(4.5)^2(36 - 1)+(5.2)^2(42 - 1)}{115 - 3} = \frac{2452.43}{112} = 21.897.$$

(d) The F statistic is calculated as

$$F = \frac{MSG}{MSE} = \frac{10.016}{21.897} = 0.457.$$

The numerator has $I - 1 = 3 - 1$ degrees of freedom and the denominator has $N - I = 115 - 3 = 112$ degrees of freedom. Going to Table D with degrees of freedom (2, 100) we see that the P-value is greater than 0.1. When the F-value is below 1, there is no need to go to the table – there is no evidence of a difference in weight loss after six months among the three treatments.

Exercise 25.35

State: To study the rate of decay of polyester in landfills, a researcher buried strips of polyester in soil for different lengths of time, then dug up the strips and measured the force required to break them. Breaking strength was chosen as it is easy to measure and should be a good indicator of decay with lower breaking strength indicating greater decay. Fifteen strips were buried in well drained soil and 5 strips, chosen at random, were dug up after 2, 4, and 8 weeks. The breaking strengths in pounds follow.

Breaking Strength-2 Weeks	Breaking Strength-4 Weeks	Breaking Strength-8 Weeks
118	130	122
126	120	136
126	114	128
120	126	146
129	128	140

Formulate: The ratio of the largest to the smallest standard deviations is

$$\frac{\text{largest sample standard deviation}}{\text{smallest sample standard deviation}} = \frac{9.53}{4.60} = 2.07$$

which is slightly larger than 2. The rule of thumb is conservative, and with equal sample sizes in the three groups many statisticians would proceed with the ANOVA in this situation. The one-way ANOVA will be used to determine if there is evidence of a difference in mean breaking strength among polyester strips buried in soil for 2, 4, or 8 weeks.

Solve: The following output is from MINITAB. The first thing to notice is that the sample mean breaking strengths are 123.80 for the two-week treatment, 123.60 for the four-week treatment and 134.40 for the eight-week treatment. The P-value is 0.056, which provides slight evidence of a difference in breaking strengths for the three groups. However, the ANOVA does not demonstrate that polyester is losing strength over the time period studied.

One-way ANOVA: Breaking Strength versus Weeks

```
Source  DF      SS     MS     F      P
Weeks    2   381.7  190.9  3.70  0.056
Error   12   619.2   51.6
Total   14  1000.9

S = 7.183   R-Sq = 38.14%   R-Sq(adj) = 27.83%
```

```
                              Individual 95% CIs For Mean Based on
                              Pooled StDev
Level   N    Mean    StDev    ---+---------+---------+---------+------
2       5    123.80   4.60    (---------*---------)
4       5    123.60   6.54    (---------*---------)
8       5    134.40   9.53                      (---------*---------)
                              ---+---------+---------+---------+------
                              119.0     126.0     133.0     140.0
```

Pooled StDev = 7.18

Conclude: Since it seems unlikely that polyester could be getting stronger over time, we would consider that the difference in sample means, although somewhat large, can be explained by chance despite the *P*-value of 0.056. The explanation would be that decay did not occur over 8 weeks and some of the stronger strips ended up in the 8-week group just by chance. Since the study may not have been carried out over a long enough period of time to see an effect of time on breaking strength, further experiments over longer time periods may need to be run.

CHAPTER 26

NONPARAMETRIC TESTS

OVERVIEW

Many of the statistical procedures described in previous chapters assume that samples are drawn from normal populations. **Nonparametric tests** do not require any specific form for the distributions of the populations from which the samples are drawn. Many nonparametric tests are **rank tests**; that is, they are based on the **ranks** of the observations rather than on the observations themselves. When ranking the observations from smallest to largest, tied observations receive the average of their ranks.

The **Wilcoxon rank sum test** compares two distributions. The objective is to determine if one distribution has systematically larger values than the other. The observations are ranked, and the **Wilcoxon rank sum statistic** W is the sum of the ranks of one of the samples. The Wilcoxon rank sum test can be used in place of the **two-sample t test** when samples are small or the populations are far from normal.

Exact P-values for the Wilcoxon rank sum test require special tables and are produced by some statistical software. However, many statistical software packages give only approximate P-values based on a normal approximation, typically with a continuity correction. Many packages also make an adjustment in the normal approximation when there are ties in the ranks.

The **Wilcoxon signed rank test** is a nonparametric test for matched pairs. It tests the null hypothesis that there is no systematic difference between the observations within a pair against the alternative that one observation tends to be larger.

The test is based on the **Wilcoxon signed rank statistic** W^+, which provides another example of a nonparametric test using ranks. The absolute values of the differences between matched pairs of observations are ranked and the sum of the ranks of the positive (or negative) differences gives the value of W^+. The **matched pairs t test** is an alternative test that assumes a normal distribution for the differences.

P-values for the signed rank test can be found in special tables of the distribution or a normal approximation to the distribution of W^+. Some software computes the exact P-value and other software uses the normal approximation, typically with a continuity correction. Many packages make an adjustment in the normal approximation when there are ties in the ranks.

The **Kruskal-Wallis test** is the nonparametric test for the **one-way analysis of variance** setting. In comparing several populations, it tests the null hypothesis that the distribution of the response variable is the same in all groups and the alternative hypothesis that some groups have distributions of the response variable that are systematically larger than others.

The **Kruskal-Wallis statistic** H compares the average ranks received for the different samples. If the alternative is true, some should be larger than others. Computationally, it essentially arises from performing the usual one-way ANOVA to the ranks of the observations rather than the observations themselves.

P-values for the Kruskal-Wallis test can be found in special tables of the distribution or a chi-square approximation to the distribution of H. When the sample sizes are not too small, the distribution of H for comparing I populations has approximately a chi-square distribution with $I-1$ degrees of freedom. Some software computes the exact P-value and other software uses the chi-square approximation, typically with an adjustment in the chi-square approximation when there are ties in the ranks.

GUIDED SOLUTIONS

Exercise 26.13

KEY CONCEPTS: Ranking data, two-sample problem, Wilcoxon rank sum test

(a) Order the observations from smallest to largest. Use a different color for or underline observations in the supplemented group to make it easier to determine the ranks assigned to each group.

(b) Suppose the first sample is the supplemented group and the second sample is the control group. The choice of which sample we call the first sample and which we call the second sample is arbitrary. However, the Wilcoxon rank sum test is the sum of the ranks of the first sample, and the formulas for the mean and variance of W distinguish between the sample sizes for the first and the second samples. Use the ranks of the supplemented group to compute the value of W.

$$W =$$

(c) What are the values of n_1, n_2, and N? Use these values to evaluate the mean and standard deviation of W according to the formulas that follow.

$$\mu_W = \frac{n_1(N+1)}{2} =$$

$$\sigma_W = \sqrt{\frac{n_1 n_2(N+1)}{12}} =$$

Use the mean and standard deviation to compute the standardized rank sum statistic.

$$z = \frac{W - \mu_W}{\sigma_W} =$$

What kind of values would W have if the alternative were true? Use the normal approximation to find the approximate P-value. If you have access to software or tables to evaluate the exact P-value, compare it with the approximation.

P-value =

What are your conclusions?

Exercise 26.25

KEY CONCEPTS: Matched pairs, Wilcoxon signed rank statistic

(a) First give the null and alternative hypotheses. If the cola loses sweetness, what will be the sign of the sweetness loss (sweetness before storage minus sweetness after storage)?

H_0:

H_a:

To compute the Wilcoxon signed rank statistic, order the absolute values of the differences and rank them. When there are ties, be careful computing the ranks. In any tied group of observations, each observation should each receive the average rank for the group. (Note that the negative observations are in bold and italics.) The ranks of the two smallest absolute values are given to help get you started. Now fill in the remaining ranks.

Absolute values	Ranks
0.4	1.5
0.4	1.5
0.7	
1.1	
1.2	
1.3	
2.0	
2.0	
2.2	
2.3	

To see how the ranks are computed, the 0.4's would get ranks 1 and 2, so their average rank is 1.5. The 0.7 would get rank 3 and so on. If W^+ is the sum of the ranks of the positive observations, compute the value of W^+.

$$W^+ =$$

Evaluate the mean and standard deviation of W^+ according to the following formulas.

$$\mu_{W^+} = \frac{n(n+1)}{4} =$$

$$\sigma_{W^+} = \sqrt{\frac{n(n+1)(2n+1)}{24}} =$$

Now use the mean and standard deviation to compute the standardized rank sum statistic.

$$z = \frac{W^+ - \mu_{W^+}}{\sigma_{W^+}} =$$

Do you expect W^+ to be small or large if the alternative is true? Use the normal approximation to find the approximate P-value.

What are your conclusions?

(b) How do the P-values from the Wilcoxon signed rank test and the one-sample t test compare?

For the one-sample t test, give the null and alternative hypotheses.

H_0:
H_a:

What are the assumptions for each of the procedures?

Exercise 26.49

KEY CONCEPTS: One-way ANOVA, Kruskal-Wallis statistic

We are going to use the Kruskal-Wallis test to determine if nematodes in soil affect plant growth. First give the null and alternative hypotheses for the Kruskal-Wallis test.

H_0:

H_a:

To compute the Kruskal-Wallis test statistic, the 16 observations are first arranged in increasing order as follows, where we have kept track of the group for each observation. Fill in the ranks. Remember that there is one tied observation.

Growth	3.2	4.6	5.0	5.3	5.4	5.8	7.4
Group	10000	5000	5000	10000	5000	10000	5000
Rank							

Growth	7.5	8.2	9.1	9.2	10.8	11.1	11.1
Group	10000	1000	0	0	0	1000	1000
Rank							

Growth	11.3	13.5
Group	1000	0
Rank		

Fill in the following table, which gives the ranks for each of the nematode groups and the sum of ranks for each group.

Nematodes	Ranks	Sum of ranks
0		
1000		
5000		
10000		

Use the sum of ranks for the four groups to evaluate the Kruskal-Wallis statistic. What are the numerical values of n_i and N in the formula?

$$H = \frac{12}{N(N+1)} \sum \frac{R_i^2}{n_i} - 3(N+1) =$$

The value of H is compared with critical values in Table E for a chi-square distribution with $I-1$ degrees of freedom, where I is the number of groups. What is the P-value and what do you conclude?

COMPLETE SOLUTIONS

Exercise 26.13

(a) First the observations are ordered from smallest to largest. The observations given in bold are from the supplemented group.

Observations	Ranks
−1.2	1
2.3	2
4.6	3.5
4.6	3.5
5.4	5
6.0	6
7.7	7.5
7.7	7.5
11.3	9.5
11.3	9.5
11.4	11
15.5	12
16.5	13

(b) The Wilcoxon rank sum statistic is

$$W = 5 + 7.5 + 9.5 + 9.5 + 11 + 12 + 13 = 67.5$$

(c) The sample sizes are $n_1 = 7$, $n_2 = 6$, and $N = 13$. The values for the mean and variance are

$$\mu_W = \frac{n_1(N+1)}{2} = \frac{7(13)}{2} = 45.5$$

and

$$\sigma_W = \sqrt{\frac{n_1 n_2 (N+1)}{12}} = \sqrt{\frac{(7)(6)(13)}{12}} = 6.745$$

and the standardized rank sum statistic W is

$$z = \frac{W - \mu_W}{\sigma_W} = \frac{67.5 - 45.5}{6.745} = 3.26$$

Since we would expect W to have large values if the alternative were true, the approximate P-value is $P(Z \geq 3.26) = 0.0006$. There is very strong evidence that the supplemented birds miss the peak by more days than the control birds.

Exercise 26.25

(a) The null and alternative hypotheses are

H_0: median = 0

H_a: median > 0

The ranks of the absolute values are

Absolute values	Ranks
0.4	1.5
0.4	1.5
0.7	3
1.1	4
1.2	5
1.3	6
2.0	7.5
2.0	7.5
2.2	9
2.3	10

The Wilcoxon signed rank statistic is

$$W^+ = 1.5 + 3 + 4 + 5 + 7.5 + 7.5 + 9 + 10 = 47.5$$

The values for the mean and variance are

$$\mu_{W^+} = \frac{n(n+1)}{4} = \frac{10(11)}{4} = 27.5$$

and

$$\sigma_{W^+} = \sqrt{\frac{n(n+1)(2n+1)}{24}} = \sqrt{\frac{(10)(11)(21)}{24}} = 9.811$$

and the standardized signed rank statistic W^+ is

$$\frac{W^+ - \mu_{W^+}}{\sigma_{W^+}} \geq \frac{47.5 - 27.5}{9.811} = 2.04$$

If the cola lost sweetness, we would expect the differences (before storage − after storage) to be positive. Thus the ranks of the positive observations should be large and we would expect the value of the statistic W^+ to be large when the alternative hypothesis is true. The approximate P-value is $P(Z \geq 2.04) = 0.021$. We conclude that the cola does lose sweetness in storage.

The output from the Minitab computer package on the next page gives a similar result. Minitab includes a correction to the standard deviation in the normal approximation to account for the ties in the ranks, so. the P-value given by Minitab is slightly different than the one we obtained.

Wilcoxon Signed Rank Test

```
TEST OF MEDIAN = 0.000000 VERSUS MEDIAN G.T.  0.000000

                N FOR    WILCOXON             ESTIMATED
            N    TEST    STATISTIC  P-VALUE    MEDIAN
Loss       10     10        47.5     0.023      1.150
```

(b) The conclusions are the same and the P-values are also quite similar. The one-sample t test hypotheses are

$$H_0: \mu = 0$$

$$H_a: \mu > 0$$

Both tests assume that the tasters in the study are a simple random sample of all tasters. The one-sample t test also assumes that the (before storage) − (after storage) sweetness differences are normally distributed.

Exercise 26.49

The null and alternative hypotheses for the Kruskal-Wallis test are

H_0: seedling growths have the same distribution in all groups

H_a: seedling growths are systematically higher in some groups than in others

When the distributions have the same shape, the null hypothesis for the Kruskal-Wallis is that the median growth in all groups are equal, and the alternative hypothesis is that not all four medians are equal.

The computations required for the Kruskal-Wallis test statistic follow.

```
Growth    3.2     4.6     5.0     5.3     5.4     5.8     7.4
Group   10000    5000    5000   10000    5000   10000    5000
Rank        1       2       3       4       5       6       7

Growth    7.5     8.2     9.1     9.2    10.8    11.1    11.1
Group   10000    1000       0       0       0    1000    1000
Rank        8       9      10      11      12    13.5    13.5

Growth   11.3    13.5
Group    1000       0
Rank       15      16
```

Nematodes	Ranks	Sum of ranks
0	10, 11, 12, 16	49
1000	9, 13.5, 13.5, 15	51
5000	2, 3, 5, 7	17
10000	1, 4, 6, 8	19

$$H = \frac{12}{N(N+1)} \sum \frac{R_i^2}{n_i} - 3(N+1) = \frac{12}{16(16+1)} \left(\frac{49^2}{4} + \frac{51^2}{4} + \frac{17^2}{4} + \frac{19^2}{4} \right) - 3(16+1) = 11.34$$

Since $I = 4$ groups, the sampling distribution of H is approximately chi-square with $4 - 1 = 3$ degrees of freedom. From Table E we see that the P-value is approximately 0.01. There is strong evidence of a difference in seedling growth between the four groups.

The MINITAB software gives the following output when doing the Kruskal-Wallis test. The medians, average ranks (in place of sums of ranks), H statistic and P-value are given. The H statistic with an adjustment for ties in the ranks is also given.

Kruskal-Wallis Test

```
LEVEL      NOBS      MEDIAN   AVE. RANK
    1         4      10.000        12.3
    2         4      11.100        12.8
    3         4       5.200         4.2
    4         4       5.550         4.7
OVERALL      16                     8.5

H = 11.34   d.f. = 3   p = 0.010
H = 11.35   d.f. = 3   p = 0.010 (adjusted for ties)
```

CHAPTER 27

STATISTICAL PROCESS CONTROL

OVERVIEW

In practice, work is often organized into a chain of activities that lead to some result. A chain of activities that turns inputs into outputs is called a **process**. A process can be described by a **flowchart,** which is a picture of the stages of a process. A **cause-and-effect diagram,** which displays the logical relationships between the inputs and output of a process, is also useful for describing and understanding a process.

All processes have variation. If the pattern of variation is stable over time, the process is said to be in statistical control. In this case, the sources of variation are called **common causes.** If the pattern is disrupted by some unusual event, **special cause** variation is added to the common cause variation. **Control charts** are statistical plots intended to warn when a process is disrupted or **out of control.**

Standard **3σ control charts** plot the values of some statistic Q for regular samples from the process against the time order in which the samples were collected. The **center line** of the chart is at the mean of Q. The **control limits** lie three standard deviations of Q above (the **upper control limit**) and below (the **lower control limit**) the center line. A point outside the control limits is an **out-of-control signal.** For **process monitoring** of a process that has been in control, the mean and standard deviations used to establish the center line and control limits are based on past data and are updated regularly.

When we measure some quantitative characteristic of a process, we use $\bar{x}$ and s **charts** for process control. The $\bar{x}$ chart plots the sample means of samples of size n from the process and the s chart the sample standard deviations. The s chart monitors variation within individual samples from the process. If the s chart is in control, the $\bar{x}$ chart monitors variation from sample to sample. To interpret charts, always look first at the s chart.

For a process that is in control with mean μ and standard deviation σ, the 3σ $\bar{x}$ chart based on samples of size n has center line and control limits

$$\text{CL} = \mu \quad \text{UCL} = \mu + 3\frac{\sigma}{\sqrt{n}} \quad \text{LCL} = \mu - 3\frac{\sigma}{\sqrt{n}}$$

The 3σ s chart has control limits

$$\text{UCL} = (c_4 + 2c_5)\sigma = B_6\sigma \quad \text{LCL} = (c_4 - 2c_5)\sigma = B_5\sigma$$

and the values of c_4, c_5, B_5, and B_6 can be found in Table 24.3 in your textbook for n from 2 to 10.

An *R* **chart** based on the range of observations in a sample is often used in place of an *s* chart. We will rely on software to produce these charts. Formulas can be found in books on quality control. $\bar{x}$ and *R* charts are interpreted the same way as $\bar{x}$ and *s* charts.

It is common to use various **out-of-control signals** in addition to "one point outside the control limits." In particular, a **runs signal** (nine consecutive points above the center line or nine consecutive points below the center line) for an $\bar{x}$ chart allows one to respond more quickly to a gradual drift in the process center.

We almost never know the mean μ and standard deviation σ of a process. They must be estimated from past data. We estimate μ by the mean $\bar{\bar{x}}$ of the observed sample means $\bar{x}$. We estimate σ by

$$\hat{\sigma} = \frac{\bar{s}}{c_4}$$

where $\bar{s}$ is the mean of the observed sample standard deviations. **Control charts based on past data** are used at the **chart setup** stage for a process that may not be in control. Start with control limits calculated from the same past data that you are plotting. Beginning with the *s* chart, narrow the limits as you find special causes, and remove the points influenced by these causes. When the remaining points are in control, use the resulting limits to monitor the process.

Statistical process control maintains quality more economically than inspecting the final output of a process. Samples that are **rational subgroups** (subgroups that capture the features of the process in which we are interested) are important to effective control charts. A process in control is stable, so we can predict its behavior. If individual measurements have a normal distribution, we can give the **natural tolerances.**

A process is **capable** if it can meet or exceed the requirements placed on it. Control (stability over time) does not in itself improve capability. Remember that control describes the internal state of the process, whereas capability relates the state of the process to external specifications.

There are control charts for several different types of process measurements. One important type is the *p* **chart**, a control chart based on plotting sample proportions $\hat{p}$ from regular samples from a process against the order in which the samples were taken. We estimate the process proportion *p* of "successes" by

$$\bar{p} = \frac{\text{total number of successes in past samples}}{\text{total number of opportunities in these samples}}$$

and then the control limits for a *p* chart for future samples of size *n* are

$$\text{UCL} = \bar{p} + 3\sqrt{\frac{\bar{p}(1-\bar{p})}{n}} \quad \text{CL} = \bar{p} \quad \text{LCL} = \bar{p} - 3\sqrt{\frac{\bar{p}(1-\bar{p})}{n}}$$

The interpretation of *p* charts is very similar to that of $\bar{x}$ charts. The out-of-control signals used are also the same as for $\bar{x}$ charts.

GUIDED SOLUTIONS

Exercise 27.1

KEY CONCEPTS: Flowcharts and cause-and-effect diagrams

For this exercise, it is important to choose a process that you know well so that you can describe it carefully and recognize those factors that affect the process. Use the space provided for your flowchart and cause-and-effect diagram.

Exercise 27.4

KEY CONCEPTS: Pareto charts

What percent of total losses do these 9 DRGs account for?

　　　Sum of percent losses =

Use the axes to make your Pareto chart.

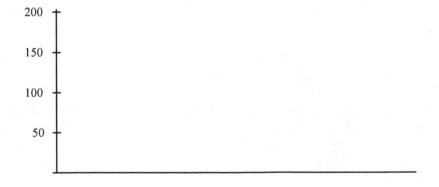

Which DRGs should the hospital study first when attempting to reduce its losses?

Exercise 27.7

KEY CONCEPTS: Common causes

Refer to Exercise 27.1 in this Study Guide. For a process you know well, what are some common sources of variation in the process?

What are some special causes that might drive the process out of control?

Exercise 27.15

KEY CONCEPTS: $\bar{x}$ and s charts

For the first two samples in Figure 27.10 of your textbook compute $\bar{x}$ and s.

Sample 1

$\bar{x} =$

$s =$

Sample 2

$\bar{x} =$

$s =$

If you have access to statistical software, use the software to make your $\bar{x}$ and s charts. Otherwise, to make the $\bar{x}$ chart, compute

$$UCL = \mu + 3\frac{\sigma}{\sqrt{n}} =$$

$$CL = \mu =$$

$$LCL = \mu - 3\frac{\sigma}{\sqrt{n}} =$$

Plot the UCL, CL, LCL, and values of $\bar{x}$ for all 18 samples.

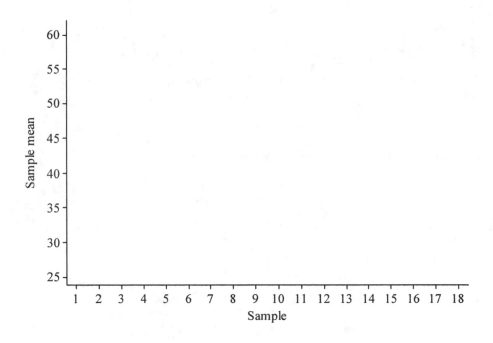

To make the s chart, compute

$$\text{UCL} = B_6 \sigma =$$

$$\text{CL} = c_4 \sigma =$$

$$\text{LCL} = B_5 \sigma =$$

Plot the UCL, CL, LCL, and the values of s for all 18 samples in the chart that follows. How would you describe the state of the process?

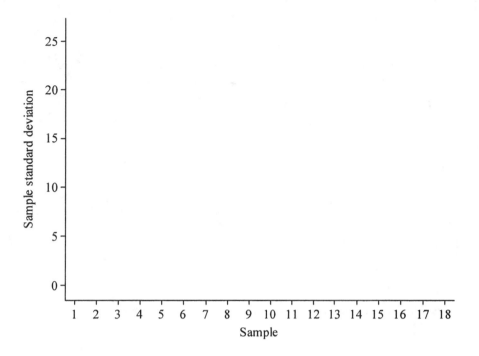

Exercise 27.20

KEY CONCEPTS: $\bar{x}$ and s control charts using past data

(a) From the values of $\bar{x}$ and s in Table 27.1 of your textbook, compute (by hand, a calculator, or using software)

$\bar{\bar{x}}$ = mean of the 20 values of $\bar{x}$ =

$\bar{s}$ = mean of the 20 values of s =

Hence we estimate μ to be

$\hat{\mu} = \bar{\bar{x}} =$

and we estimate σ to be

$$\hat{\sigma} = \frac{\bar{s}}{c_4} =$$

(b) Look at the s chart in Figure 27.7 of your textbook. What patterns do you see that might suggest that the process σ may now be less than 43 mV?

Exercise 27.29

KEY CONCEPTS: Natural tolerances

The natural tolerances are $\mu \pm 3\sigma$. We do not know μ and σ, so we must estimate them from the data. We remove sample 5 from the data. Based on the remaining 17 samples, estimate

$\bar{\bar{x}}$ = mean of the 17 values of $\bar{x}$ =

$\bar{s}$ = mean of the 17 values of s =

Hence we estimate μ to be

$$\hat{\mu} = \bar{\bar{x}} =$$

and we estimate σ to be

$$\hat{\sigma} = \frac{\bar{s}}{c_4} =$$

Based on these estimates, the natural tolerances for the distance between the holes are

$$\hat{\mu} \pm 3\hat{\sigma} =$$

Exercise 27.30

KEY CONCEPTS: Capability

Refer to Exercise 24.29 in this Study Guide. Based on the 17 samples that were in control, we see that estimates of μ and σ are $\hat{\mu} = 43.41$ and $\hat{\sigma} = 12.39$. We therefore assume that distances between holes vary from meter to meter according to an $N(43.41, 12.39)$ distribution. Use normal probability calculations to find the probability that the distance x between holes in a randomly selected meter is between 54 ± 10 (i.e., between 44 and 64). Refer to Chapter 3 of your textbook if you have forgotten how to do normal probability calculations.

$$P(44 < x < 64) =$$

We conclude that about what percent of meters meet specifications?

Exercise 27.34

KEY CONCEPTS: p charts

To find the appropriate center line and control limits, we must first compute $\bar{p}$. The total number of opportunities for missing or deformed rivets is just the total number of rivets, because each rivet has the possibility of being missing or deformed. The number of "successes" in past samples is just the missing or deformed rivets in the recent data. What are these values? Now estimate $\bar{p}$.

$$\bar{p} = \frac{\text{total number of successes in past samples}}{\text{total number of opportunities in these samples}} =$$

The next wing contains $n = 1070$ rivets, and the control limits for a p chart for future samples of size $n = 1070$ are

$$\text{UCL} = \bar{p} + 3\sqrt{\frac{\bar{p}(1 - \bar{p})}{n}} =$$

$$\text{CL} = \bar{p} =$$

$$\text{LCL} = \bar{p} - 3\sqrt{\frac{\bar{p}(1 - \bar{p})}{n}} =$$

COMPLETE SOLUTIONS

Exercise 27.1

We take as our example the process of making a cup of coffee. A possible flowchart and cause-and-effect diagram of the process follow.

Flowchart   Cause-and-Effect Diagram

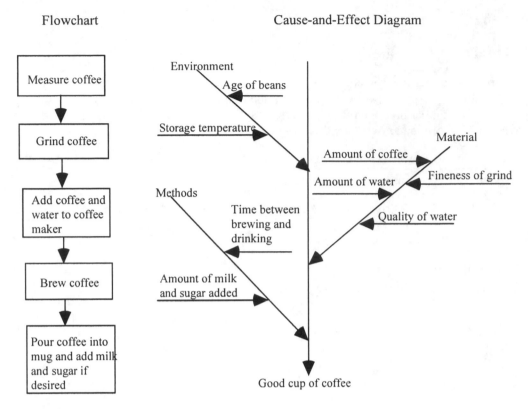

Exercise 27.4

Adding the percents listed, the percent of total losses that these 9 DRGs account for is 80.5%. A Pareto chart of losses by DRG follows.

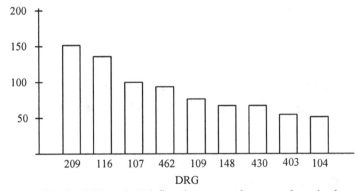

The hospital ought to study DRGs 209 and 116 first in attempting to reduce its losses. These are the two DRGs with the largest percent losses and combined account for nearly 30% of all losses.

Exercise 27.7

In Exercise 27.1 of this Study Guide, we described the process of making a good cup of coffee. Some sources of common-cause variation are variation in how long the coffee has been stored and the conditions under which it has been stored, variation in the measured amount of coffee used, variation in how finely the coffee is ground, variation in the amount of water added to the coffee maker, variation in the length of time the coffee sits between when it has finished brewing and when it is drunk, and variation in the amount of milk and/or sugar added.

Some special causes that might at times drive the process out of control would be a bad batch of coffee beans, a serious mismeasurement of the amount of coffee used or the amount of water used, a malfunction of the coffee maker or a power outage, interruptions that result in the coffee sitting a long time before it is drunk, and the use of milk that has gone bad.

Exercise 27.15

We compute $\bar{x}$ and s for the first two samples:

First sample: $\bar{x} = 48$, $s = 8.94$ Second sample: $\bar{x} = 46$, $s = 13.03$

To make the $\bar{x}$ chart, we note that

$$\text{UCL} = \mu + 3\frac{\sigma}{\sqrt{n}} = 43 + 3\frac{12.74}{\sqrt{5}} = 43 + 17.09 = 60.09$$

$$\text{CL} = \mu = 43$$

$$\text{LCL} = \mu - 3\frac{\sigma}{\sqrt{n}} = 43 - 3\frac{12.74}{\sqrt{5}} = 43 - 17.09 = 25.91$$

resulting in the chart that follows.

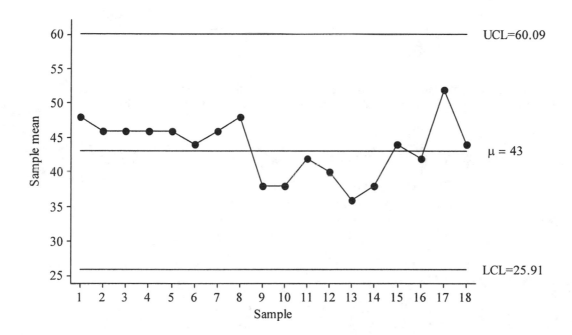

To make the s chart, we note that

$$UCL = B_6\sigma = 1.964(12.74) = 25.02$$

$$CL = c_4\sigma = 0.9400(12.74)) = 11.98$$

$$LCL = B_5\sigma = 0(12.74)) = 0$$

resulting in the chart that follows.

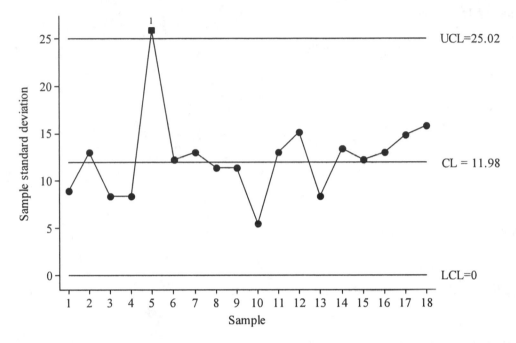

The s chart shows a lack of control at sample point 5, but otherwise neither chart shows a lack of control. We would want to find out what happened at sample 5 to cause a lack of control in the s chart.

Exercise 27.20

(a) From the values of $\bar{x}$ and s in Table 27.1 of the textbook, we compute (using software)

$$\bar{\bar{x}} = \text{mean of the 20 values of } \bar{x} = 275.065$$
$$\bar{s} = \text{mean of the 20 values of } s = 34.55$$

Hence we estimate μ to be

$$\hat{\mu} = \bar{\bar{x}} = 275.065$$

and we estimate σ to be (using the fact that the samples are each of size $n = 4$ and according to Table 24.3 of the textbook, $c_4 = 0.9213$)

$$\hat{\sigma} = \frac{\bar{s}}{c_4} = \frac{34.55}{0.9213} = 37.5$$

(b) If we look at the s chart in Figure 27.7 of the textbook we see that most of the points lie below 40 (and more than half of the points below 40 lie well below 40), while of the points above 40, all but one (sample 12) are only slightly larger than 40. The s chart suggests that typical values of s are below 40, which is consistent with the estimate of σ in part (a).

Exercise 27.29

The natural tolerances are $\mu \pm 3\sigma$. We do not know μ and σ, so we must estimate them from the data. We remove sample 5 from the data. Based on the remaining 17 samples, we find

$$\bar{\bar{x}} = \text{mean of the 17 values of } \bar{x} = 43.41$$
$$\bar{s} = \text{mean of the 17 values of } s = 11.65$$

Hence we estimate μ to be

$$\hat{\mu} = \bar{\bar{x}} = 43.41$$

and we estimate σ to be (using the fact that the samples are each of size $n = 5$ and according to Table 24.3 of the textbook, $c_4 = 0.9400$)

$$\hat{\sigma} = \frac{\bar{s}}{c_4} = \frac{11.65}{0.9400} = 12.39$$

Based on these estimates, the natural tolerances for the distance between the holes are

$$\hat{\mu} \pm 3\,\hat{\sigma} = 43.41 \pm 3(12.39) = 43.41 \pm 37.17 \text{ or } 6.24 \text{ to } 80.58$$

Exercise 27.30

Based on the 17 samples that were in control, we saw in Exercise 27.29 in this Study Guide that estimates of μ and σ are $\hat{\mu} = 43.41$ and $\hat{\sigma} = 12.39$. We therefore assume that distances between holes vary from meter to meter according to an $N(43.41,12.39)$ distribution. The probability that the distance x between holes in a randomly selected meter is between 54 ± 10 (i.e., between 44 and 64) is thus

$$P(44 < x < 64) = P(\frac{44 - 43.41}{12.39} < \frac{x - 43.41}{12.39} < \frac{64 - 43.41}{12.39}) = P(0.05 < Z < 1.66)$$

$$= P(Z < 1.66) - P(Z < 0.05) = 0.9515 - 0.5199 = 0.4316$$

We conclude that about 43.16% of meters meet specifications.

Exercise 27.34

The total number of opportunities for missing or deformed rivets is just the total number of 34700 rivets, because each rivet has the possibility of being missing or deformed. The number of "successes" in past samples is just the 208 missing or deformed rivets in the recent data. We therefore estimate the process proportion p of "successes" from the recent data by

$$\bar{p} = \frac{\text{total number of successes in past samples}}{\text{total number of opportunities in these samples}} = \frac{208}{34700} = 0.00599$$

The next wing contains $n = 1070$ rivets, and the control limits for a p chart for future samples of size $n = 1070$ are

$$\text{UCL} = \bar{p} + 3\sqrt{\frac{\bar{p}(1-\bar{p})}{n}} = 0.00599 + 3\sqrt{\frac{0.00599(1-0.00599)}{1070}} = 0.00599 + 0.00708 = 0.01307$$

$$\text{CL} = \bar{p} = 0.00599$$

$$\text{LCL} = \bar{p} - 3\sqrt{\frac{\bar{p}(1-\bar{p})}{n}} = 0.00599 - 3\sqrt{\frac{0.00599(1-0.00599)}{1070}} = 0.00599 - 0.00708 = 0$$

Note that in the LCL, we set negative values to 0 because a proportion can never be less than 0.

CHAPTER 28

MULTIPLE REGRESSION

OVERVIEW

Multiple linear regression extends the techniques of simple linear regression to situations involving $p > 1$ explanatory variables $x_1, x_2, \ldots, x_p$. The data consist of the values of the response y and the p explanatory variables for n individuals or cases. Data analysis begins by examining the distribution of the variables individually and then drawing scatterplots to explore the relationships between the variables.

The mean response μ_y for a **multiple regression model** based on p explanatory variables $x_1, x_2, \ldots, x_p$ is

$$\mu_y = \beta_0 + \beta_1 x_1 + \beta_2 x_2 + \ldots + \beta_p x_p$$

The multiple regression equation predicts the value of the response y as a linear function of the explanatory variables

$$\hat{y}_i = b_0 + b_1 x_{i1} + b_2 x_{i2} + \ldots + b_p x_{ip}$$

where the coefficients b_i are estimated using the method of least squares. The variability of the responses about the multiple regression equation is measured in terms of the **regression standard error** s,

$$s = \frac{\sum e_i^2}{n - p - 1}$$

where the e_i are the **residuals**:

$$e_i = y_i - \hat{y}_i$$

The regression standard error s has $n - p - 1$ degrees of freedom. The **distribution of the residuals** should be examined and the residuals should be plotted against each of the p explanatory variables. In practice, the b's and s are calculated using statistical software.

A special case of the multiple linear regression model is fitting separate regression lines to two sets of data. Fitting the lines is done using an **indicator variable** to show from which data set an observation comes and using an **interaction** term to allow for different slopes.

The **ANOVA table** for a multiple regression is analogous to that in simple linear regression. It gives the sum of squares, the mean squares, and the degrees of freedom for regression and residual sources of variation. The ANOVA F is the regression mean square (*MSM*) divided by the residual mean square

(*MSE*) and is used to test the hypothesis H_0: $\beta_1 = \beta_2 = \ldots = \beta_p = 0$. Under H_0, this statistic has an $F(p, n - p - 1)$ distribution.

The **squared multiple correlation** can be written as the ratio of model to total variation, namely,

$$R^2 = \text{SSM/SST}$$

and is interpreted as the proportion of the variability in the response variable y that is explained by the explanatory variables $x_1, x_2, \ldots, x_p$. in the multiple regression.

A **level C confidence interval** for β_j is

$$b_j \pm t^* \, \text{SE}_{b_j}$$

where t^* is the upper $(1 - C)/2$ critical value for the $t(n - p - 1)$ distribution. SE_{b_j} is the standard error of b_j and in practice is computed using statistical software.

The **test of the hypothesis H_0: $\beta_j = 0$** is based on the *t* **statistic**

$$t = \frac{b_j}{\text{SE}_{b_j}}$$

with *P*-values computed from the $t(n - p - 1)$ distribution. In practice, statistical software is used to carry out these tests.

In multiple regression, interpretation of these confidence intervals and tests depends on the particular explanatory variables in the multiple regression model. The estimate of β_j represents the effect of the explanatory variable x_j when it is added to a model already containing the other explanatory variables. The test of H_0: $\beta_j = 0$ tells us if the improvement in the ability of our model to predict the response y by adding x_j to a model already containing the other explanatory variables is statistically significant. It does not tell us if x_j would be useful for predicting the response in multiple regression models with a different collection of explanatory variables.

Confidence intervals for the mean response μ_y have the form

$$\hat{y} \pm t^* \, \text{SE}_{\hat{\mu}}$$

Prediction intervals for an individual future response y have the form

$$\hat{y} \pm t^* \, \text{SE}_{\hat{y}}$$

where t^* is the critical value for the $t(n-p-1)$ density curve. $\text{SE}_{\hat{\mu}}$ and $\text{SE}_{\hat{y}}$ can be computed using statistical software. In practice, both confidence intervals for μ_y and prediction intervals for an individual future observation are computed using statistical software.

GUIDED SOLUTIONS

Exercise 28.15

KEY CONCEPTS: Regression with indicator variables

(a) Review Exercise 4.7 if you have forgotten how to make a scatterplot using separate symbols for a categorical variable. If you are not using software to make the plot, use the axis that follow for your plot. To get you started, we have plotted the first point for men (using the symbol x) and the first point for women (using the symbol o).

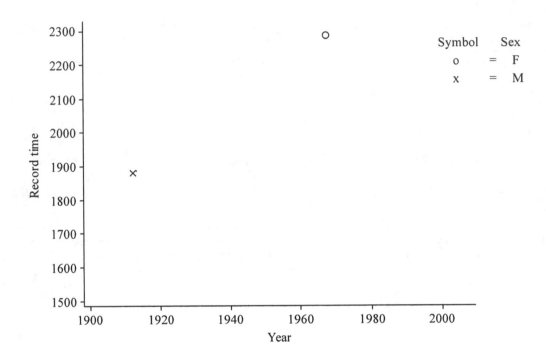

How would you describe the pattern for each sex? Do the points for each sex tend to follow a straight line or some curved relation?

How would you describe the progress of men and women?

(b) To fit a model with two regression lines, one for men and one for women, you will need to create an indicator variable for sex. To allow for lines of different slopes you will also need to create a variable representing the interaction between sex and year. After doing so, use software to fit a multiple regression model with year, the indicator variable for sex, and the variable representing the interaction between sex and year as predictors.

Estimated model with two regression lines:

Estimated regression line for men:

Estimated regression line for women:

(c) Do the data appear to support any of these claims? If you know recent world record times for men and women you might see if the rate of improvement for women has changed and if the difference in record times for men and women has become negligible.

Exercise 28.25

KEY CONCEPTS: Multiple linear regression, R^2, F test, t tests

The following Minitab output for the regression of weight on length and width can be used to help answer parts (a) through (d). You should run the regression with the software you are using in your course to become familiar with the format of the output. Although, the regression output should be similar to the Minitab output, there may be slight variations in the names for some of the quantities.

Regression Analysis: Weight versus Length, Width

```
The regression equation is
Weight = - 579 + 14.3 Length + 113 Width

Predictor      Coef   SE Coef       T      P
Constant    -578.76     43.67  -13.25  0.000
Length       14.307     5.659    2.53  0.014
Width        113.50     30.26    3.75  0.000
```

```
S = 88.6760    R-Sq = 93.7%    R-Sq(adj) = 93.5%


Analysis of Variance

Source          DF      SS        MS        F       P
Regression       2   6229332   3114666   396.09   0.000
Residual Error  53    416762      7863
Total           55   6646094
```

(a) As part of the Minitab output, the formula for the estimated regression equation is provided. If you are using a different software package, you may need to use the estimated coefficients to write the equation. Use the information in the output to give the estimated multiple regression equation

$$\hat{y} =$$

(b) Which regression quantity measures the amount of variation in the response explained by the model? It is included in the output.

Amount of variation in weight explained by the model in (a) =

(c) The null and alternative hypotheses tested by the ANOVA F test are

H_0:

H_a:

Does a test of these hypotheses answer the question posed? Both the test statistic and P-value are included with the output.

(d) The individual t tests that β_1 and β_2 are significantly different from zero are included in the output. What do you conclude from them?

The following Minitab output for the regression of weight on length, and width and their interaction can be used to help answer parts (e) through (h). When using your software, you will need to first create a new column for the product of length and width and then include this variable "Interaction" in the model.

Regression Analysis: Weight versus Length, Width, Interaction

```
The regression equation is
Weight = 114 - 3.48 Length - 94.6 Width + 5.24 Interaction


Predictor        Coef   SE Coef       T       P
Constant       113.93     58.78    1.94   0.058
Length         -3.483     3.152   -1.10   0.274
Width          -94.63     22.30   -4.24   0.000
Interaction    5.2412    0.4131   12.69   0.000


S = 44.2381    R-Sq = 98.5%    R-Sq(adj) = 98.4%


Analysis of Variance

Source            DF        SS        MS        F       P
Regression         3   6544330   2181443  1114.68   0.000
Residual Error    52    101765      1957
Total             55   6646094
```

(e) As part of the Minitab output, the formula for the estimated regression equation is provided. If you are using a different software package, you may need to use the estimated coefficients to write the equation. Use the information in the output to give the estimated multiple regression equation.

$\hat{y} =$

(f) Which regression quantity measures the amount of variation in the response explained by the model? It is included in the output.

Amount of variation in weight explained by the model in (e) =

(g) The null and alternative hypotheses tested by the ANOVA F test are

H_0:

H_a:

Does a test of these hypotheses answer the question posed? Both the test statistic and P-value are included with the output.

(h) When the explanatory variables are correlated, the estimated coefficients change as well as their individual t statistics. Since the interaction term is the product of length and width, it is correlated with both length and width. Describe how the individual t statistics change when the interaction term is added.

Exercise 28.27

KEY CONCEPTS: Confidence intervals for the mean, prediction intervals

Confidence intervals for the mean and prediction intervals require specifying a list of values for all the explanatory variables in the model. You are asked to obtain these intervals for the tenth perch. What are the values of the explanatory variables for this perch?

Length =

Width =

Interaction =

Software packages differ in how they obtain confidence and prediction intervals. In some packages, such as SAS, if you ask for these intervals they are automatically produced for the explanatory variables at every observation. In other packages, such as Minitab, you must specify the explanatory variables for which you want confidence and prediction intervals. You should learn how to obtain these intervals with the software you are using for this course. The Minitab output follows.

```
Predicted Values for New Observations

New
Obs    Fit   SE Fit      95% CI           95% PI
  1   84.02   10.41   (63.13, 104.91)   (-7.18, 175.21)


Values of Predictors for New Observations

New
Obs  Length  Width  Interaction
  1    21.0   2.80         58.8
```

What *t* distribution was used to obtain these intervals?

Interpret both intervals.

Exercise 28.29

KEY CONCEPTS: Residual plots

Recall that the conditions for inference require agreement between the observed and predicted values (residuals centered about a horizontal line through 0), constant variance (the residuals look like an unstructured band of points centered around a horizontal line through 0), and normality (absence of outliers in the residual plot).

Do you see any problems in either of the plots?

COMPLETE SOLUTIONS

Exercise 28.15
(a)

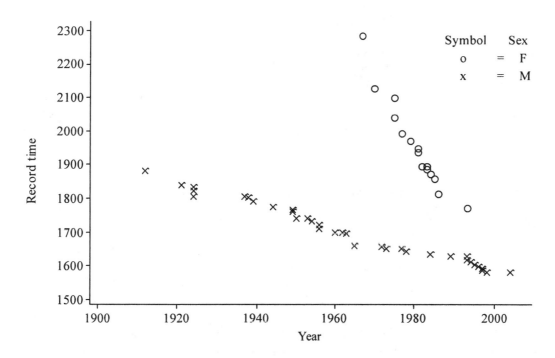

For both men and women the points tend to follow a straight line. Thus, we would describe the pattern as linear and decreasing.

The progress for women appears better than for men. Times for women have been decreasing more rapidly than the times for men.

(b) We used the variable "ind" as the indicator variable for sex. Ind = 1 represents males and ind = 0 females. The interaction term is the variable ind*year. Here are the Minitab results of the fitted lines.

```
The regression equation is
Record time  = 41373 - 19.9 year - 33247 ind + 16.6 ind*year


Predictor         Coef  SE Coef         T       P
Constant         41373     1713     24.15   0.000
year          -19.9046   0.8652    -23.01   0.000
ind             -33247     1734    -19.17   0.000
ind*year       16.6326   0.8759     18.99   0.000


S = 21.1736   R-Sq = 98.3%   R-Sq(adj) = 98.2%
```

Based on these results we have

Estimated model with two regression lines: record time = 41373 – 19.9×year – 33247×ind + 16.6×ind*year

Estimated regression line for men: record time = (41373 – 33247) + (–19.9 + 16.6) ×year
= 8126 – 3.3×year

Estimated regression line for women: record time = 41373 – 19.9×year

(c) The improvement for women (at least for the data we have) is more rapid than for men. The difference in record times is decreasing, but it is not negligible. One is tempted to extrapolate the data and argue that the difference will become negligible as time passes. However, extrapolation is dangerous. If one looks at the trend in the two lines, one might expect that the difference in world record times for men and women to be negligible by about the year 2000 and perhaps that women would have a faster world record time than men after 2000. This is not the case, at least as of 2006.

Exercise 28.25

(a) The estimated multiple regression equation can be read off the Minitab output as

```
    Weight = - 579 + 14.3 Length + 113 Width
```

Alternatively, the estimated equation can be obtained from the coefficient column in the output reading the values for the constant, length coefficient and width coefficient to yield

$$\hat{y} = -578.76 + 14.307 \text{ length} + 113.50 \text{ width}$$

(b) The quantity R^2 measures the variability explained by the model, or in this case the variation in the weight of a perch explained by its length and width. R^2 is given on the Minitab output as

```
            R-Sq = 93.7%.
```

Thus almost 94% of the variation is the weight of a perch is explained by its length and width.

(c) The null and alternative hypotheses tested by the ANOVA F test are

$$H_0: \beta_1 = \beta_2 = 0$$

$$H_a: \text{ at least one of } \beta_1 \text{ and } \beta_2 \text{ is not } 0$$

where β_1 is the regression coefficient of "length" and β_2 is the regression coefficient of "width" in our multiple linear regression model. From the output, $F = 396.09$ and the P-value $= 0.000$. We reject the null hypothesis, so at least one of length or width is helpful in predicting the weight of the perch.

(d) The individual t statistic for length is $t = 2.53$ with P-value $= 0.014$. This indicates that length helps explain the weight of the perch even after we allow width to explain the weight. The individual t statistic for length is $t = 3.75$ with P-value $= 0.000$. This indicates that width helps explain the weight of the perch even after we allow length to explain the weight. The combined information from these indicates that both variables should be used to explain the weight of the perch.

(e) The estimated multiple regression equation can be read off the Minitab output as

```
Weight = 114 - 3.48 Length - 94.6 Width + 5.24 Interaction
```

Alternatively, the estimated equation can be obtained from the coefficient column for the constant, length coefficient, width coefficient, and interaction coefficient to yield

$$\hat{y} = 114 - 3.483 \text{ length} - 94.63 \text{ width} + 5.2412 \text{ interaction}$$

(f) The quantity R^2 measures the variability explained by the model, or in this case the variation in the weight of a perch explained by its length and width. R^2 is given on the Minitab output as

```
R-Sq = 98.5%
```

Thus almost 99% of the variation is the weight of a perch is explained by its length, width and their interaction.

(g) The null and alternative hypotheses tested by the ANOVA F test are

$$H_0: \beta_1 = \beta_2 = \beta_3 = 0$$

$$H_a: \text{ at least one of } \beta_1, \beta_2 \text{ and } \beta_3 \text{ is not } 0$$

where β_1 is the regression coefficient of "Length", β_2 is the regression coefficient of "Width" and β_3 is the regression coefficient of interaction in our multiple linear regression model. From the output, $F = 1114.68$ and the P-value $= 0.000$. We reject the null hypothesis, so at least one of length, width or their interaction is helpful in predicting the weight of the perch.

(h) When interaction is added, the individual t for length changes from 2.53 to -1.10 and the P-value changes from 0.014 to 0.274. Not only is the statistical significance affected, but the sign of the estimated coefficient changes from positive to negative. The individual t for width changes from 3.75 to −4.24, with the statistical significance unaffected. However, the sign of the estimated coefficient has changed. The meaning of the coefficients for length and width change with the addition of the interaction term, so these changes are not surprising. In this example, we can see that the relationship between the response y and any one explanatory variable can change greatly depending on what other explanatory variables are present in the model.

Exercise 28.27

A $t(n - p - 1) = t(56 - 3 - 1)$, or t distribution with 52 degrees of freedom was used to obtain the intervals. We are 95% confident that the average weight of all fish with a length of 21 and a width of 2.8 is between 63.13 and 104.91. We are 95% confident that a new fish caught with this length and width will have a weight no larger that 175.21. We can take the lower endpoint as zero since negative weights are not possible.

Exercise 28.29

The first plot has somewhat of a funnel shape, with variability increasing as the variable Purchase12 increases. Otherwise, there are no other obvious problems (no outliers and the residuals appear to be centered about 0).

The second plot also has a funnel shape with the variability decreasing as Recency increases. The points appear to be centered about 0 and there are no outliers in the vertical direction, but the point at the far right is an outlier in the horizontal directions. This may be an influential observation.

We conclude that the condition that appears to be most suspect is the constant variance assumption. There is nothing in either plot that suggests there are serious problems with the assumption of normality or independence. We might be concerned about the effect of the possible influential observation on the fitted model (especially the effect on the impact of the predictor variable Recency).